太庙享殿前组群南立面图
South elevation of the front groups of the Sacrificial Hall of the Ancestral Temple

太庙大戟门前组群南立面图
South elevation of the front groups of the Daji Gate of the Ancestral Temple

太庙建筑组群西立面图

West elevation of the groups of the Sacrificial Hall of the Ancestral Temple

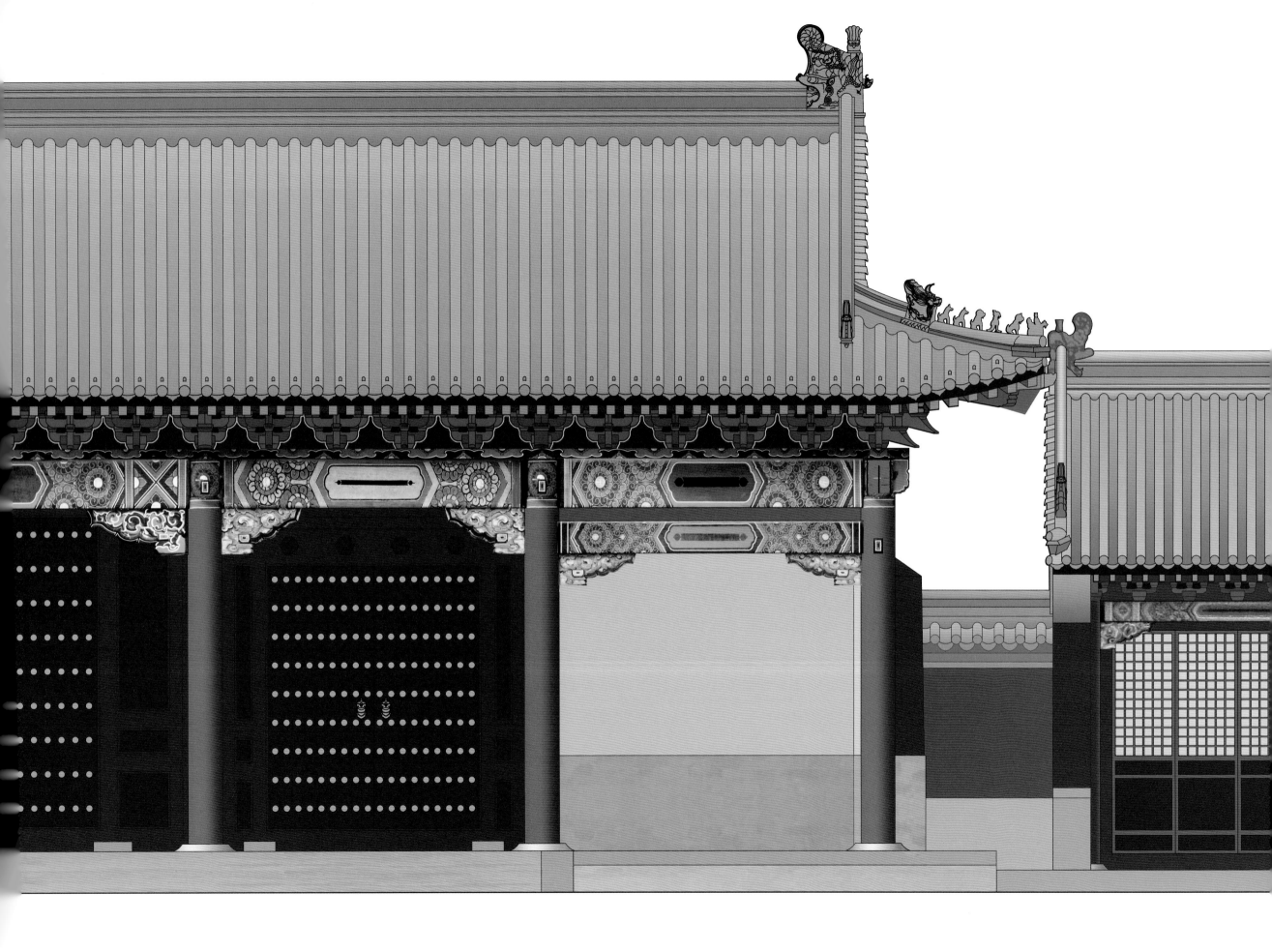

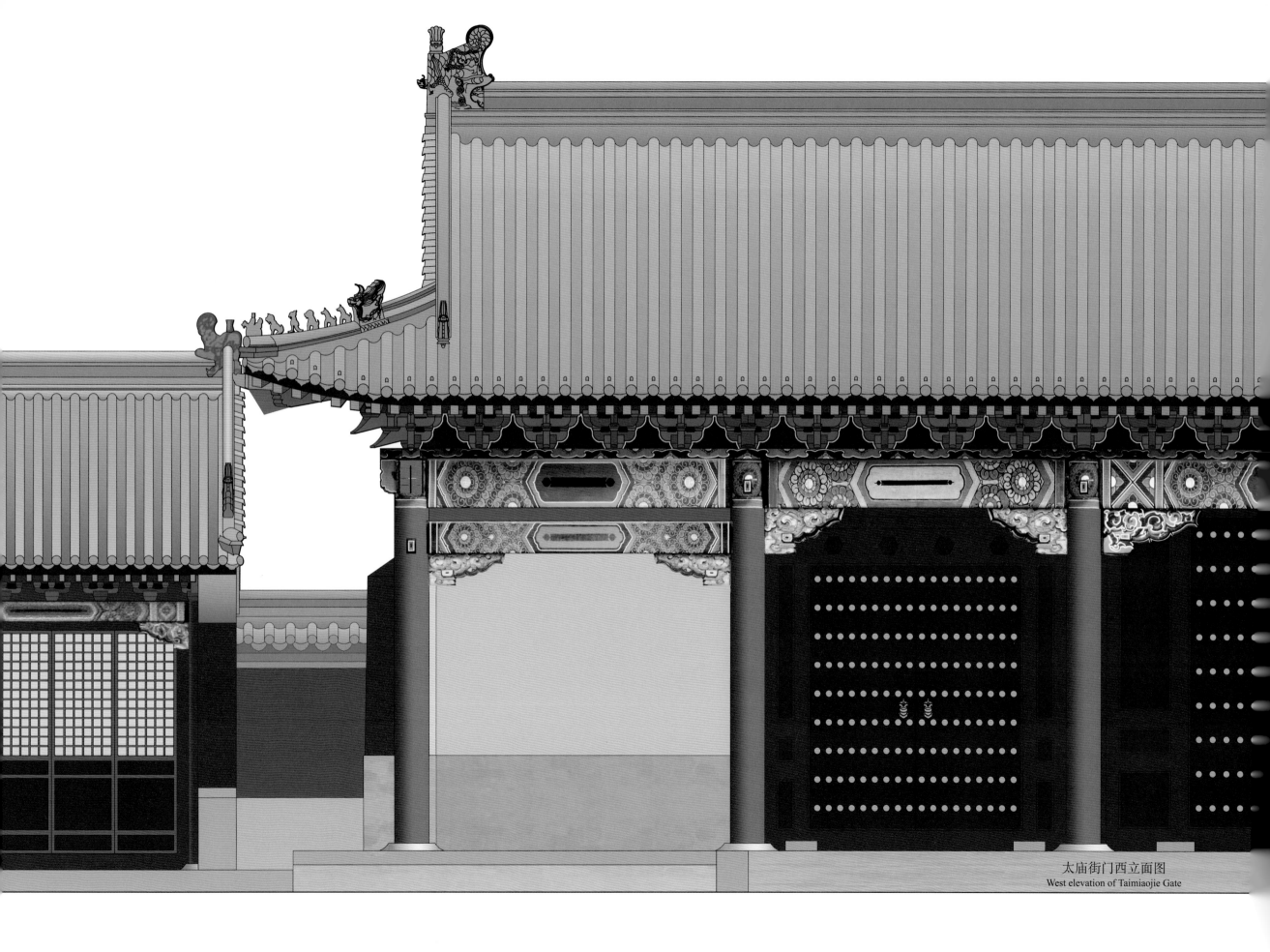

太庙街门西立面图
West elevation of Taimiaojie Gate

中国古建筑测绘大系·坛庙建筑

『十二五』国家重点图书出版规划项目

国家出版基金项目

太庙和社稷坛

天津大学建筑学院　北京市劳动人民文化宫　北京市中山公园管理处　合作编写

王其亨　主编　何捷　王其亨　编著

中国建筑工业出版社

Traditional Chinese Architecture Surveying and
Mapping Series:
Temples Architecture

# THE ANCESTRAL TEMPLE AND
# THE ALTAR OF LAND AND GRAIN

Compiled by School of Architecture, Tianjin University
Beijing Working People's Culture Palace
Administration Office of Beijing Zhongshan Park
Chief Edited by WANG Qiheng
Edited by HE Jie, WANG Qiheng

China Architecture & Building Press

# Contents

**Location of the Ancestral Temple:** Southeast of Forbidden City in Beijing

**Year of initial construction:** 1420

**Occupied area:** 16.9 hectares

**Governing body:** Beijing Working People's Cultural Palace

**Surveying and mapping unit:** School of Architecture, Tianjin University

**Years of surveying and mapping:** 1997-1998

**Location of the Altar of Land and Grain:** Southwest of the Forbidden City in Beijing

**Year of initial construction:** 1420

**Occupied Area:** 23.8 hectares

**Governing body:** Administration Office of Beijing Zhongshan Park

**Surveying and mapping unit:** School of Architecture, Tianjin University

**Year of Surveying and mapping:** 1996

太庙地址：北京紫禁城东南方

始建年代：一四二〇年

占地面积：十六点九公顷

主管单位：北京市劳动人民文化宫

测绘单位：天津大学建筑学院

测绘时间：一九九七年至一九九八年

社稷坛地址：北京紫禁城东南方

始建年代：一四二〇年

占地面积：二十三点八公顷

主管单位：北京中山公园管理处

测绘单位：天津大学建筑学院

测绘时间：一九九六年

# Introduction

The Ancestral Temple and the Altar of Land and Grain are located in the southern part of the imperial city on either side of the central axis of Beijing (Fig.1, Fig.2). They follow the established practice within Chinese capital city construction of placing the Ancestral Temple in the east and the Altar of Land and Grain in the west ("temple in east and altar in west" for short). The construction of the Ancestral Temple and the Altar of Land and Grain was initiated in the Ming Dynasty (1368-1644) when Emperor Chengzu (1402-1424) built the palaces in Beijing. Following their completion in 1420, the temple and the altar continued to be used until the end of the Qing Dynasty (1644-1911). The Ancestral Temple served as a ceremonial architectural complex in which the reigning emperor worshipped his ancestors. Covering an area of about 16.9 hectares, it consists of a central area (including the Daji Gate, the three major halls and their side halls), the inner buildings and an outer peripheral area. The Altar of Land and Grain was where the royal family offered sacrifices to the God of Land and the God of Grain. Including the additional sections allocated to the modern Zhongshan Park, the altar now covers an area of about 23.8 hectares and consists of the central architectural ensemble, the outbuildings within the two altar walls, as well as buildings that were relocated to the site or newly constructed in modern times. In 1988, both the Ancestral Temple and the Altar of Land and Grain were designated as key cultural relic units under national protection by the State Council.

導　言

太庙和社稷坛作为北京都城的『左祖右社』，位于宫城南部，分列于中轴线东西两侧（图一，图二）。太庙与社稷坛始于明成祖营建北京宫殿，建成于永乐十八年（一四二〇年），一直沿用到清末。太庙是明清两代皇帝祭祀祖先的礼制建筑群，占地面积约十六点九公顷（按全国重点文物保护单位规划保护范围，不计世庙），主要由大戟门、三大殿及配殿组成的中心区，头道宫墙及内部建筑和外部周边区域组成，其中文物建筑基址占地面积约一万四千七百一十平方米。社稷坛是皇家祭祀土神和五谷神的场所，包括近现代中山公园突破原社稷坛扩建和额外划拨的地段，面积约二十三点八公顷。由中心建筑群及两道坛墙以内的附属建筑，以及近现代新建、迁建的建筑组成。太庙和社稷坛均于一九八八年由国务院公布为第三批全国重点文物保护单位。

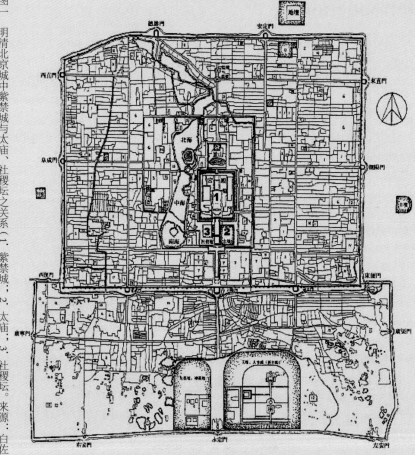

图一　明清北京城中紫禁城与太庙、社稷坛之关系（1. 紫禁城；2. 太庙；3. 社稷坛。来源：自佐民、邵俊仪，中国美术全集·建筑艺术编6·坛庙建筑[M]. 北京：中国建筑工业出版社，1994 ）

图二　北京太庙、社稷坛卫星影像图（来源：谷歌地球）

002

Fig.1　Relationship between the Forbidden City and the Ancestral Temple and the Altar of Land and Grain in Beijing during the Ming and Qing dynasties (1. Forbidden City; 2. Ancestral Temple; 3. Altar of Land and Grain).　Source: BAI zuomin. *Complete Works of Chinese Architectural Art · Altars and Shrines Buildings[M]*. Beijing: China Architecture & Building Press, 1994)

Fig.2　Satellite images of Beijing Ancestral Temple, Altar of Land and Grain (Source:cited from Google Earth)

# I.Construction of the Ancestral Temple and the Altar of Land and Grain during the Ming and Qing Dynasties

## (I) Construction of the Ancestral Temple and the Altar of Land and Grain in Nanjing during the Ming Dynasty

The Ancestral Temple and the Altar of Land and Grain in Beijing are located on the east and west sides of the Meridian Gate, respectively. Their design follows the system of traditional Chinese capital construction, which places the Ancestral Temple in the east and the Altar of Land and Grain in the west. Many records on capital planning can be found in early literature. [1] Emperor Wen (581-604) of the Sui Dynasty first incorporated the prescription of "temple in the east and altar in the west" into the construction of his capital city, Daxing (now Xi'an, Shaanxi). Serving as the capital of the Sui (581-618) and Tang (618-907) dynasties, this city established a model for capital city planning in terms of the layout of its palaces and altars. [2] After a recession of capital construction in the Song (960-1279), Liao (916-1125), Jin (1115-1234) and Yuan (1279-1368) dynasties, the process of city planning based on classical models was reestablished in the Ming Dynasty under the Hongwu Emperor (1368-1398). While the altar and shrine ceremonies of the Zhou (1045-256 BC), Han (206 BC-220 AD) and Tang dynasties were restored during this time, the layout of the imperial palaces and urban structures—including the Ancestral Temple and the Altar of Land and Grain—was reworked for Hongwu's Central Capital (now Fengyang, Anhui) and Southern Capital (now Nanjing, Jiangsu Province). The layout of these buildings determined under Hongwu formed the foundation for the subsequent construction of Beijing.

In December 1366, ZHU Yuanzhang and his ministers agreed to start construction of the Ancestral Temple, the Altar of Land and Grain, and the palaces in the following year. [3] Meanwhile, ZHU Yuanzhang ordered ritual officials and Confucian officials to investigate the sacrifical rituals of previous emperors, and to draft rules and regulations on altars and temples of the current dynasty, as well as sacrificial rites for the reference of later generations. In August 1367, the Altar of Land and Grain was completed in the southwest of Nanjing Imperial City, facing north, with the Altar of Land and the Altar of Grain separately enshrined within the city walls (Fig.3). [4][5] In September, the Ancestral Temple was completed, with the "temples located in southeast of the imperial palace, all facing south" [6] as "resting places of the capital palace" that held the memorial tablets in four ancestral temples [7] (Fig.4).

The suspended construction of the Central Capital in April 1375 depressed ZHU Yuanzhang, who then ordered the revision of the ancestral temple system of Nanjing in July 1375. According to the

# 一、明清北京太庙与社稷坛的建设

## （一）明代南京的太庙与社稷坛建设

北京太庙与社稷坛分别位于紫禁城午门南方之左右两侧，是按照古代『左祖右社』的都城结构布局进行规划设计的。这一都城空间规划策略在早期文献中有大量记述[1]。在都城建设实践中，隋文帝大兴城开始将『左祖右社』置于皇城之内。经过隋唐两京的营建，都城宫殿与坛庙格局基本定型[2]。经过了宋辽金元的都城营造衰退期之后，明代洪武朝稽古创制，在恢复周、汉、唐之坛庙祀典仪轨的同时，在中都和南京有机组织包括太庙与社稷坛在内的宫城与城市空间，为后世北京的城市营建打下了坚实基础。

元至正二十六年（一三六六年）十二月，朱元璋与群臣议定次年开始营建宗庙、社稷与宫室[3]。同时在祀典方面，朱元璋命礼官和儒臣稽考故帝王之祭祀、拟定国朝坛庙之规制以及祭祀之仪轨，以垂鉴后世。吴元年（一三六七年）八月癸丑，社稷坛建成于南京宫城西南，北向，太稷分祀于同墙内的两坛[4][5]（图三）。同年九月甲戌，太庙成，『庙在宫城东南，皆南向』[6]，采用『都宫别殿』，奉安四祖庙[7]（图四）。

由于洪武八年（一三七五年）四月中都停建，无法释怀的太祖朱元璋于七月即开始改南京太庙制度，『度地阙左』[8]，重新选址于午门外左侧。洪武九年（一三七六年）十月，太庙成[9]，承袭中

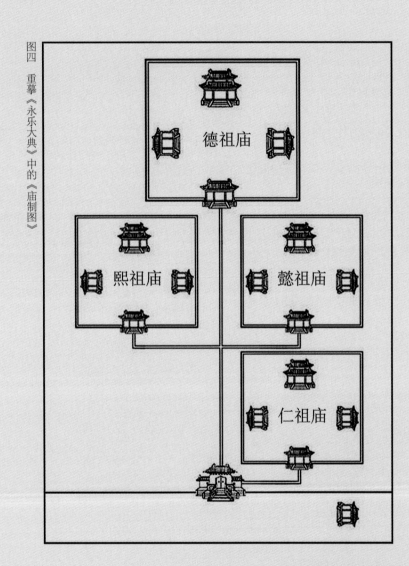

图四　重摹《永乐大典》中的《庙制图》

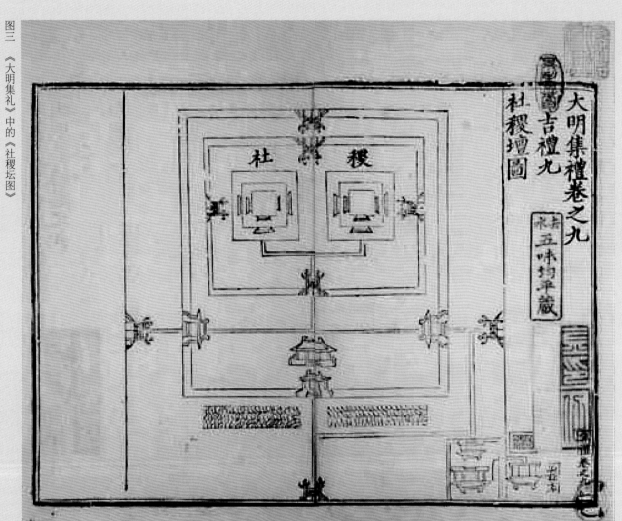

图三　《大明集礼》中的《社稷坛图》

Fig.3　*Altar of Land and Grain map* in *A Collection of Etiquettes of the Ming Dynasty*
Fig.4　*Recopied temple drawing* in the *Yongle Canon*

official record, "Land was measured for excavation in the east",or, in other words, the Ancestral Temple site was relocated to the east side of the Meridian Gate. ⑧ In October 1376, the Ancestral Temple was completed.⑨ It changed the temple system in the Central Capital which featured "different rooms in the same hall."⑩ to the "hall in the front and bedroom in the rear" layout with east and west side rooms.⑪This layout became the blueprint of the Ancestral Temple in Beijing during its initial construction (Fig.5). In Aug. 1377, ZHU Yuanzhang and his ritual officials investigated the ancient ritual systems of "God of Land and God of Grain" and built a new Altar of Land and Grain on the west side out of the Meridian Gate opposite the Ancestral Temple, which better accorded to the ideal "temple in east and altar in west" layout (Fig.6). On the first day of the tenth lunar month, the new Altar of Land and Grain was completed. The Goulong and Zhouqi memorial tablets were removed, and the memorial tablet of Emperor Renzu of the Ming Dynasty was installed, elevating the Altar of Land and Grain to one of the most important sacrificial sites in the capital.⑫

# (II) Initial Construction of Ancestral Temple and Altar of Land and Grain in Beijing

Emperor Chengzu of the Ming Dynasty commenced the construction of the Beijing imperial palaces in 1417. Both the Ancestral Temple and the Altar of Land and Grain were completed in Dec. 1420. Following the arrangement of their counterparts in Nanjing, the altars were located to the south of the palace, on either side of the central axis.⑬ Like the Ancestral Temple in Nanjing, the one in Bejing comprised the "different rooms in the same hall" form⑭, though adjustments were made to the shape and structure of individual buildings. The Ancestral Temple included a main hall in the front and a bedroom hall in the rear. The halls were both nine bays across, and each bay held a memorial tablet.⑮ The main hall had 15 side-rooms on each of its wings, and the bedroom hall had 5 side-rooms on each wing (Fig.7). The Altar of Land and Grain is in strict accordance with the Nanjing blueprint in 1377, but adjustments were made to its size. The Memorial Hall and the Worshiping Hall were relocated to inside the outer walls, and the shapes and structures of the four gates on the outer walls, as well as the dimensions of the individual buildings, were changed (Fig.8). ⑯

The Ancestral Temple underwent renovations and repairs in 1446⑰ and 1457⑱. On 26th, Aug. 1487 in lunar month, Emperor ZHU Jianshen died.⑲ As all the "nine temples" in the Beijing Ancestral Temple were occupied by Emperor Dezu, Emperor Yizu, Emperor Xizu, Emperor Renzong, Emperor Taizu, Emperor Taizong, Emperor Renzong, Emperor Xuanzong, and Emperor Yingzong, it became necessary to figure out where to put ZHU Jianshen's memorial tablet. In November of the same year, it was agreed that a remote ancestral hall would be built in the back of the bedroom hall.⑳ The construction of the ancestral hall began in Feb. 1491 and was completed in May 1492. From then on,

都宗庙『同堂异室』之制⑩，变为『前殿后寝』配以东西庑的格局⑪，成为后继北京始建太庙的蓝本（图五）。其后，洪武十年（一三七七年）八月癸丑，朱元璋与礼臣考证『社稷』古制，新建社稷坛于午门外之右，与太庙隔街相对，形成更为理想的『左祖右社』格局（图六）。十月丙午朔，新建社稷坛成。社、稷共为一坛，罢勾龙、周弃配位，奉仁祖配享，社稷遂升为大祀⑫。

## （二）北京太庙与社稷坛的初期建设

明成祖于永乐十五年（一四一七年）营建北京宫殿，太庙、社稷坛均建成于永乐十八年（一四二〇年）十二月，依照南京的形制，分列于『皇城内南』⑬阙之左右。北京太庙沿用南京太庙『同堂异室』制度⑭，但对单体建筑形制进行了调整。太庙前正殿，后寝殿，各九间，每间辟一室，奉藏神主⑮。各殿均有两配殿，前殿东西庑各十五间，寝殿东西庑各五间（图七）。社稷坛亦完全依照洪武十年的南京蓝本，但对社稷坛层数、高度、尺度进行了调整，祭、拜二殿的位置也移至外垣之内，同时对外围垣四门形制以及单体建筑的尺寸进行了调整（图八）。

太庙建成后，正统十一年（一四四六年）及天顺元年（一四五七年）均有修缮⑯。明成化二十三年（一四八七年）八月己丑，明宪宗朱见深宾天⑰。由于北京太庙的『九庙』已满（即德祖、懿祖、熙祖、仁祖、太祖、太宗、仁宗、宣宗、英宗），故需要议定祧庙之仪。同年十一月，议定修建祧庙于北京太庙寝殿之后，制为『九室』⑱。弘治四年（一四九一年）二月开工建设，次年五月建成，形成了享殿、寝殿与祧庙三殿的建筑格局（图九），并修建了东、西院墙以围合祧庙，通往祧

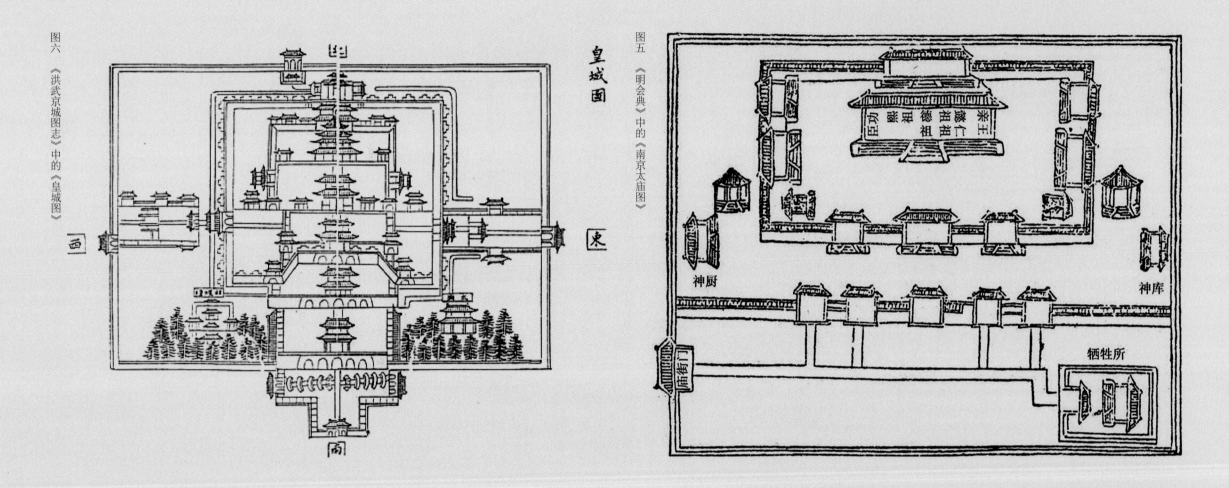

皇城图

图六 《洪武京城图志》中的《皇城图》

图五 《明会典》中的《南京太庙图》

西　東

北

南

西　東

神厨　神库

牺牲所

庙街门

Fig.5　Map of Ancestral Temple in Nanjing in the *Collected Statutes of the Ming Dynasty*
Fig.6　Imperial city map in the *Hongwu Capital City Maps*

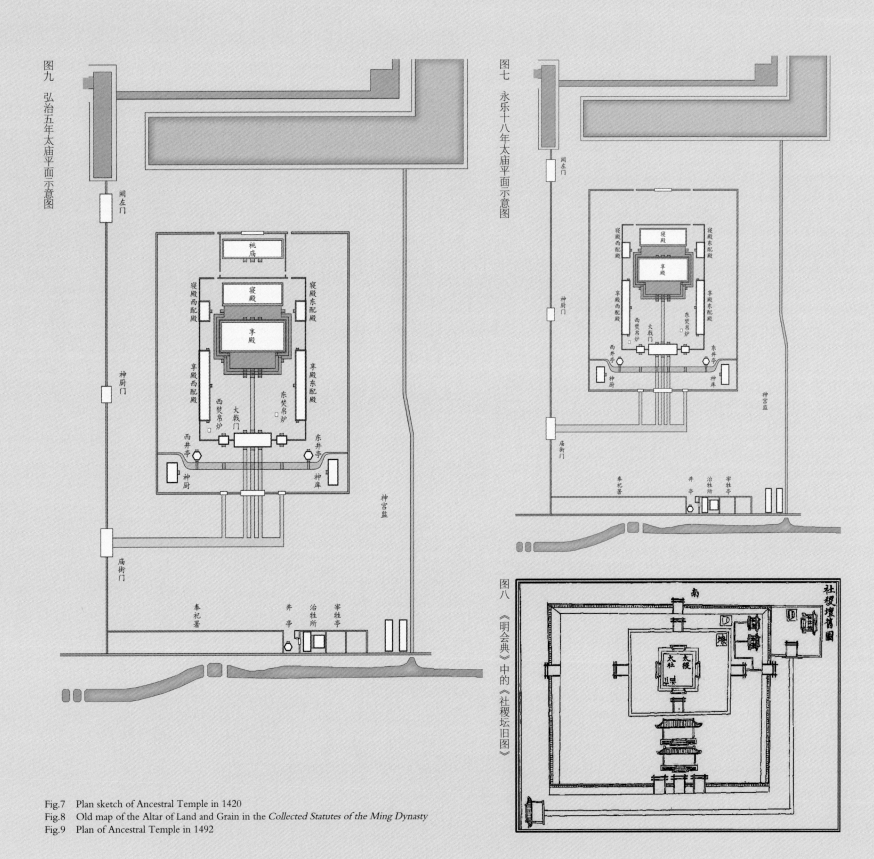

图九　弘治五年太庙平面示意图

阙左门

神厨门

庙街门

桃庙

寝殿西配殿　寝殿　寝殿东配殿

享殿西配殿　享殿东配殿

西焚帛炉　大戟门　东焚帛炉

西井亭　东井亭

神厨　神库

奉祀署

井亭　治牲所　宰牲亭

神宫监

图七　永乐十八年太庙平面示意图

阙左门

神厨门

庙街门

寝殿西配殿　寝殿　寝殿东配殿

享殿西配殿　享殿　享殿东配殿

西焚帛炉　大戟门　东焚帛炉

西井　东井

神厨　神库

奉祀署　井亭　治牲所　宰牲亭

神宫监

图八　《明会典》中的《社稷坛旧图》

社稷坛旧图

南

太社　太稷

Fig.7　Plan sketch of Ancestral Temple in 1420
Fig.8　Old map of the Altar of Land and Grain in the *Collected Statutes of the Ming Dynasty*
Fig.9　Plan of Ancestral Temple in 1492

the tripartite architectural layout of the Sacrificial Hall, the Bedroom Hall and the Remote Ancestor Hall was established (Fig.9), with the Remote Ancestor Hall enclosed by east and west courtyard walls. Paths extending through the east and west corner gates behind the Bedroom Hall led to the Remote Ancestor Hall. In 1520, the Front and Rear Halls, the east and west side rooms, the Divine Kitchens and the Divine Stores were built at the Ancestral Temple.[21] This was the last recorded renovation before the Ancestral Temple system reform in Emperor Jiajing's reign.

Following its completion in 1420, the Altar of Land and Grain underwent multiple reconstructions.[22] The system was adjusted in 1530.[23] From that point onward, the layout and buildings remained unchanged until the Qing Dynasty.

# (III) System reform in the Emperor Jiajing's reign and the construction of the Ancestral Temple

In March 1521, Emperor Wuzong, ZHU Houzhao died, and the throne was passed to his younger male cousin ZHU Houcong who later became the Emperor Jiajing. During 1521-1538, Emperor Jiajing launched a set of reforms intended to restore the clan system and show filial piety to his father. These reforms were later known as the "great debate over ceremonial systems." The altar and temple system reforms under Jiajing led to a large-scale adjustment of the Yongle period altar and temple systems, including the Ancestral Temple, suburban altars, Confucian temples, Altar of Land and Grain, Altar of the God of Silkworms, Altar of the God of Agriculture, and imperial temples. The Jiajing reforms also resulted in the construction of a great number of new altars and temples, such as the Temple of Earth, the Altar of the Sun, the Altar of the Moon, Temple of Ancient Monarchs, the Altar of the God of Silkworm and the Altar of Land and Grain.

Among all the reforms undertaken during this time, the Ancestral Temple was the most impacted in terms its layout. A main reason why Emperor Jiajing altered the layout of the Ancestral Temple was that he had moved the memorial tablet of his father—ZHU Youyuan (King Xingxian), half-brother of Emperor Xiaozong, ZHU Youtang—to the Ancestral Temple for the convenience of worshiping his father in Beijing. In 1524, the Jiajing emperor built the "Hall of Morality Observation" (Guande Dian) to worship the memorial tablet of his father.[24] In May 1525, the emperor also ordered the construction of the Shi Temple to the northeast of the Ancestral Temple for worshiping his father.[25] The temple was completed in June 1526 (Fig.10). In March 1527, the Hall of Morality Observation was completed to the east of the Hall for Ancestry Worship (Fengxian Dian), and was renamed the "Hall for Ancestry Reverence" (Chongxian Dian).[26]

## （三）嘉靖改制和太庙建设

明正德十六年（一五二一年）三月，明武宗朱厚照崩，因其无子嗣，其堂弟朱厚熜入京膺承大统，是为嘉靖帝。随后，嘉靖帝为了正宗统、孝皇考，在正德十六年至嘉靖十七年（一五二一至一五三八年）间展开了一场规模巨大、旷日持久的礼制争论和政治斗争，史称『大礼议』。经过这一时期对宗庙、郊坛、孔庙、社稷坛、先蚕坛、帝王庙等祀典与建筑制度的复议礼与更旋，嘉靖朝的坛庙改制不仅导致永乐朝北京坛庙初制的大规模调整，也选址新建了地坛、朝日坛、夕月坛、历代帝王庙、先蚕坛、帝社稷坛等大量坛庙组群。其中对太庙格局的变化影响最大。

嘉靖改制对太庙建设影响的主要原因，是嘉靖皇帝为了能在北京宗祀其皇考兴献王朱佑杬（明孝宗朱佑樘异母弟），并最终使之称宗祔庙。从嘉靖三年（一五二四年）开始，先造观德殿以奉藏兴献帝神主[24]。嘉靖四年（一五二五年）五月起再于太庙东北修建世庙供奉[25]，至嘉靖五年（一五二六年）六月，世庙工成（图十）。嘉靖六年（一五二七年）三月，完成新建观德殿于奉先殿之左方，赐名为『崇先殿』[26]。

社稷坛在永乐十八年（一四二〇年）建成之后，有明一代有多次修缮记录[22]。嘉靖九年（一五三〇年）调整祭祀制度[23]，但格局与建筑一直延续未变，为清代所继承。

庙则需经由寝殿前后之东西角门。正德十五年（一五二〇年）太庙修前后殿、两庑、神库、神厨[21]，是为嘉靖改制前最后的修缮记录。

Although King Xingxian was worshiped in the Shi Temple, Jiajing also included his father's tablet in the Ancestral Temple. In 1531, Emperor Jiajing initiated discussions on the separation of ancestors in the Ancestral Temple, after the suburban sacrifice system reform had been completed.[27] However, resistance from Jiajing's officials was so great that the reform on the temple system was not started until June 1534, after the Ancestral Temple in Nanjing had been destroyed by fire, a sign of disaster. In August of the same year, Emperor Jiajing officially announced the abandonment of the Nanjing Ancestral Temple, and declared that all the royal ancestors would henceforth be worshiped in Beijing.[28] In September, the decision to change the Beijing Ancestral Temple to follow the "separate sacrificial halls in a palace" system was again brought to the table.[29]

On 12th, Jan 1535, the construction of a total of seven temples, including the Taizong Temple and six other temples for different generations of emperors, began in the east and west areas between the inner and outer walls of the original Ancestral Temple. The original Ancestral Temple was dedicated to Emperor Taizu, ZHU Yuanzhang. A new Shi Temple was built in the southeast of the palace and renamed "Emperor Xian Temple."[30] During Sep. to Nov. 1536, the Ancestral Temple, the Taizong Temple, six other temples for different generations of emperors, and the Emperor Xian Temple were completed,[31] forming the layout of "nine temples"—eight within the palace and one (Emperor Xian's Temple) out of the palace. The main hall of the original Shi Temple was renamed the Hall of Great Deity (Jingshen Dian)[32]. In order to include the Emperor Xian Temple into the Ancestral Temple area, a north-south partition was built on the east of the ditch separating the Hall of Great Deity from the Ancestral Temple area, and three gates were built facing the stone bridge. However, because the partition was built later than the original Shi Temple, it intersected the west wall of the Shi Temple (Fig.11).[33] In Sep. 1538, the emperor honored Emperor Taizong as "Chengzu" and announced that his tablet could never be relocated. Emperor Jiajing also granted the posthumous title Ruizong to Emperor Xian so that he could be worshipped in the Ancestral Temple.[34]

In April 1541, the Beijing Ancestral Temple caught fire, and all the temples, except the Rui Temple (originally the Emperor Xian Temple) outside the palace, were destroyed.[35] The large-scale construction of altars and temples as well as palaces in Beijing over the years had put a huge strain on financial and material resources, and construction of the new Ancestral Temple did not begin until Nov. 1543.[36] The construction was completed in June 1545.[37] The "different rooms in the same hall" layout was resumed as the sacrificial rites system, the Rui Temple was dismantled, and the memorial tablet of Emperor Ruizong was installed in the Ancestral Temple. The Front Hall, the Bedroom Hall, and the east and west side rooms were rebuilt, with the Front Hall changed to have nine main bays ("rooms") and east and west sidemost bays, totaling eleven bays, while the Bedroom Hall maintained its original nine bays across the front. The east, west and south walls of the Remote Ancestor Temple

兴献王虽然在京城世庙独享，但毕竟与太庙有别。嘉靖帝为使其父称宗入庙，在更定了郊祀制度之后，于嘉靖十年（一五三一年）开始宗庙分祀之议[二十]。但由于改制阻力大，直到十三年（一五三四年）六月，借南京太庙焚毁的灾异之象重启庙制改革。同年八月，嘉靖帝正式宣布废祀南京太庙，祖宗祭祀一归京师[二十八]。九月，复议北京太庙改行『都宫别殿』的方案[二十九]。

嘉靖十四年（一五三五年）正月己亥，于太庙内、外围墙之内的东西空地开始新建太宗庙与三昭三穆共『七庙』，原正殿、寝殿仅为太祖之庙。又于都宫东南修建新世庙，改称献皇帝庙[三十]。次年（一五三六年）九月至十一月间，太庙、太宗庙、昭穆六庙与献皇帝庙相继竣工[三十一]，由此形成了都宫之内八庙与都宫之外献皇帝庙共『九庙』的格局。原世庙遂更名，正殿改为景神殿[三十二]。为了将献皇帝庙纳入宗庙区，故在沟渠以东修建了一道南北向隔墙，并将景神殿隔于宗庙区外以示分别，并修建了三座门与石桥相对。但由于此墙修建晚于原世庙，故形成了此墙与世庙西墙相错的情况（图十一）[三十三]。嘉靖十七年（一五三八年）九月，嘉靖皇帝崇太宗为『成祖』，亦为万世不祧之主，并上献皇帝庙号为『睿宗』，祔享太庙，配享明堂[三十四]。

嘉靖二十年（一五四一年）四月辛酉夜，北京太庙火灾，除都宫外的睿庙独存之外，其他群庙皆焚毁。[三十五] 由于多年来北京坛庙、宫殿的大规模建设造成的财力消耗与材料局限，直至嘉靖二十二年（一五四三年）十一月乙丑，新建太庙方正式兴工[三十六]，至嘉靖二十四年（一五四五年）六月庚申建成[三十七]。其祀典制度改回为『同堂异室』，撤除睿庙（原献皇帝庙），睿宗升祔太庙。此次重建了前殿、寝殿及东西庑，改前殿为『九室』与东西夹室的十一间格局，寝殿仍为九间；拆改祧庙东、西、南三墙，改除角道，南墙设三间琉璃花门及左右角门，增设祧庙前部月台，重建祧庙九间，添建东西庑各五间。

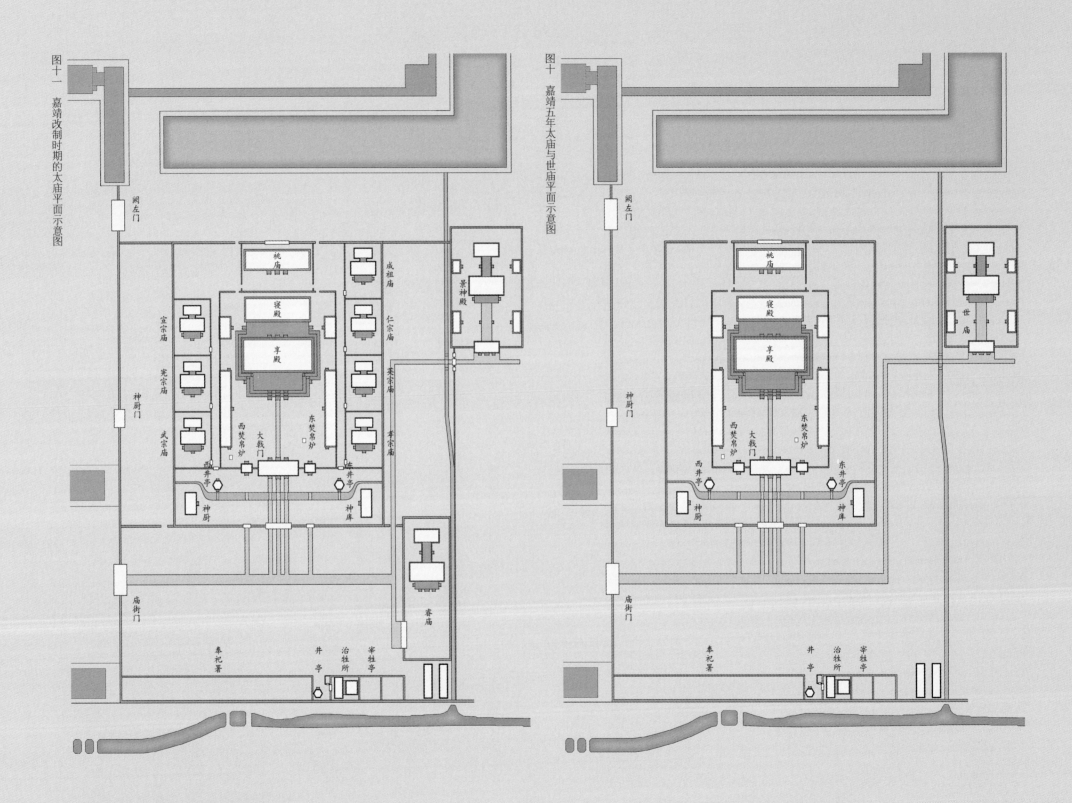

图十一　嘉靖改制时期的太庙平面示意图

阙左门

祧庙

成祖庙

寝殿

宣宗庙

仁宗庙

享殿

宪宗庙

英宗庙

神厨门

武宗庙

西焚帛炉

东焚帛炉

宪宗庙

西井亭

大戟门

库门

神厨

神库

庙街门

睿庙

奉祀署

井亭

治牲所

宰牲亭

景神殿

图十　嘉靖五年太庙与世庙平面示意图

阙左门

祧庙

寝殿

神厨门

享殿

世庙

西焚帛炉

东焚帛炉

西井亭

大戟门

东井亭

神厨

神库

庙街门

奉祀署

井亭

治牲所

宰牲亭

Fig.10　Plan of the Ancestral Temple and the Shi Temple in 1526

Fig.11　Plan sketch of Ancestral Temple during the Jiajing reform

were either dismantled or renovated. Three gates with glazed tile rooffs were added at the south wall. A terrace was newly built in the front of the Remote Ancestor Temple and the nine bays of the hall were rebuilt, with five east and five west side-rooms added. Because the fire in 1541 did not destroy the Rui Temple, many of its buildings have been handed down to us as important cultural relics today (Fig.12 and Fig.13) [38].

The Ancestral Temple system after the reform in the Jiajing period remained unchanged until the end of the Ming Dynasty and was inherited by the Qing Dynasty. The Ancestral Temple layouts depicted in the literature and drawings of the early Qing Dynasty all follow the layout of the final reconstruction in the Jiajing period (Fig.14).

## (IV) Ancestral Temple and Altar of Land and Grain in Beijing in the Qing Dynasty

The Qing rulers continued to use the Beijing Ancestral Temple and Altar of Land and Grain of the Ming Dynasty without large-scale reconstruction, except for some repairs and adjustments as needed. In May 1644, the Qing army took Beijing , and relocated their capital to Beijing from Shengjing in August of the same year. On 18th Sep. 1644, Emperor Shunzhi entered the Forbidden City through the Zhengyang Gate. On the twenty-seventh day, the memorial tablets of Emperor Taizu and Empress Taizu, as well as Emperor Taizong were moved into the Ancestral Temple. [39] In 1648, the Ancestral Temple underwent renovations[40] and its layout in the early Qing Dynasty was the same as that in the Ming Dynasty (Fig.15). During Emperor Qianlong's reign, the temple underwent multiple renovations. In particular, in the twenty-eighth year of Emperor Qianlong's reign (1763), the Inner Golden River was channeled to the originally dry moat of the Ancestral Temple and white marble balustrades and baluster columns were added.[41] The temple was also repaired in the Jiaqing and Guangxu periods,[42] but the architecture remained unchanged (Fig.16~Fig.18).

The Altar of Land and Grain was designated as the most solemn sacrificial rite in the first year of the Shunzhi Period (1644).[43] From 1740 to 1773, the altar underwent many renovations, followed by more repairs during the Xianfeng, Tongzhi and Guangxu periods.[44] However, since the altar's completion in the 1420, its plan has basically remained intact till today. This is supported by records and images in classical literature such as *Emperor Authorized Research on Archived News of Qing Capital* and *Collected Statutes of the Great Qing Dynasty* (Fig.19).

由于嘉靖二十年火灾并未蔓延到都宫之外的睿庙，戟门外的神厨库、井亭以及头道琉璃花门外的宰牲亭、治牲房、井亭等乃作为永乐朝太庙遗存沿用至今（图十二、图十三）[36]。

此后至明终世，嘉靖朝改建后的太庙建制未有改易，清代继续沿用。清代早期文献与图档中所现的均是嘉靖朝重建后太庙的格局（图十四）。

## （四）清代的北京太庙与社稷坛

清朝入关以后，直接沿用了明代北京太庙和社稷坛，其间并无大规模的改建，只是进行了适当的修缮和调整。

顺治元年（一六四四年）五月清军攻入北京城，同年八月自盛京迁都北京，九月甲辰顺治帝自正阳门入紫禁城，壬子即奉太祖帝、后和太宗神主于太庙。[39] 顺治五年（一六四八年）对太庙进行了修缮[40]，清初太庙格局与明代相比并无改变（图十五）。乾隆朝进行了多次修缮，特别是于乾隆二十八年（一七六三年）引内金水河水入原为干沟的太庙筒子河，并添建汉白玉栏板及望柱[41]。此后嘉庆、光绪等各朝均有修缮[42]，但建筑本体与组群格局并无变化（图十六～图十八）。

顺治元年（一六四四年）始确定社稷坛为大祀[43]，乾隆五年（一七四〇年）至三十八年（一七七三年）间记录有多次整修，咸丰、同治、光绪年间也有修缮记载。[44] 但北京社稷坛在永乐十八年（一四二〇年）建成后，历经明清两代建制一直完整保留至今，《钦定日下旧闻考》《大清会典》等文献记载及图像印证了这一传承（图十九）。

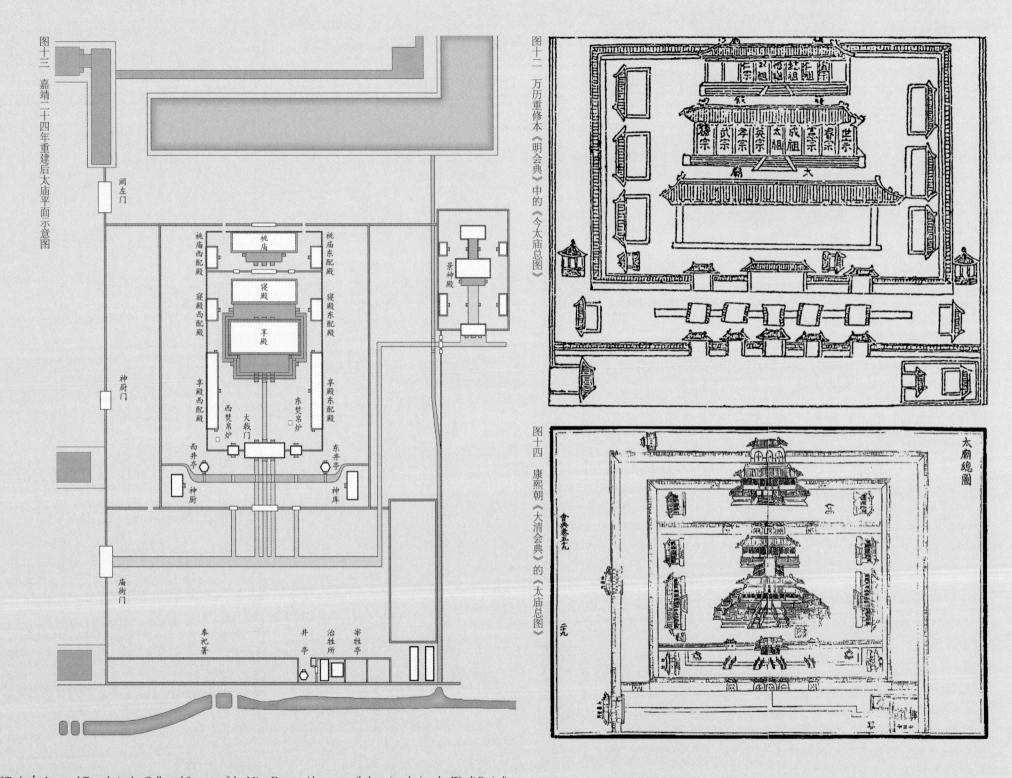

图十二 万历重修本《明会典》中的《今太庙总图》

图十三 嘉靖二十四年重建后太庙平面示意图

阙左门

神厨门

庙街门

桃庙西配殿　桃庙　桃庙东配殿

寝殿西配殿　寝殿　寝殿东配殿

享殿西配殿　享殿　享殿东配殿

西焚帛炉　大戟门　东焚帛炉

西井亭　　　　　东井亭

神厨　　　　　神库

奉祀署　　井亭　治牲亭　宰牲所

景神殿

图十四 康熙朝《大清会典》的《太庙总图》

太庙总图

会典卷五九

二九

Fig.12　*General Drawing of Today's Ancestral Temple in the Collected Statutes of the Ming Dynasty* (the recompiled version during the Wanli Period)

Fig.13　Plan sketch of Ancestral Temple after the reconstruction in 1545

Fig.14　*General Drawing of Ancestral Temple in the Collected Statutes of the Great Qing Dynasty* (compiled in the Kangxi Period)

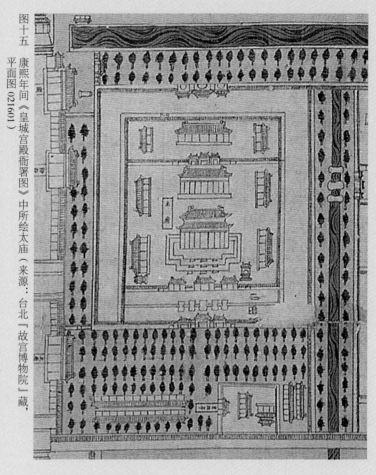

图十七　《清会典图》中的光绪年间太庙

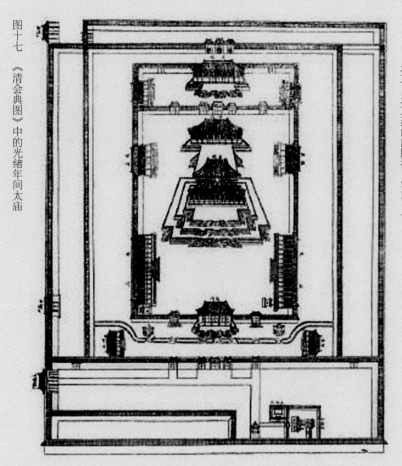

图十六　乾隆十五年《京城全图》中所绘太庙都宫部分（来源：加摹乾隆京城全图 [M].北京：北京燕山出版社，1996.）

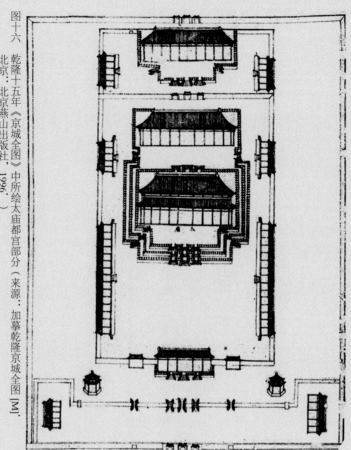

图十五　康熙年间《皇城宫殿衙署图》中所绘太庙（来源：台北『故宫博物院』藏，平面图 021601）

Fig.15　Ancestral Temple in the *Map of Palaces and Government Offices in the Imperial City* dating back to the Kangxi Period
(Source: held in the National Palace Museum in Taipei, China, Plan 021601)

Fig.16　Ancestral Temple Palace (local) in the *Full Map of the Capital City* dating back to 1750 (Source: additional copy of *Full Map of the Capital City* dating back to the Qianlong Period, Beijing: Beijing Yanshan Press, 1996)

Fig.17　The Ancestral Temple in the Guangxu Period in the *Graphs of Collected Statutes of the Qing Dynasty*.

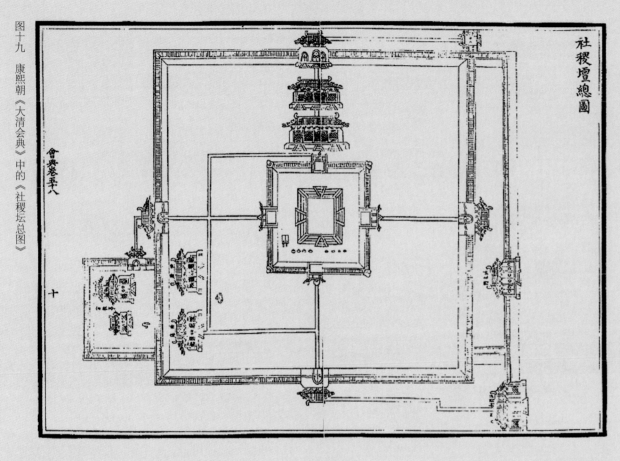

社稷壇總圖

图十九　康熙朝《大清会典》中的《社稷坛总图》

會典卷卅八

十

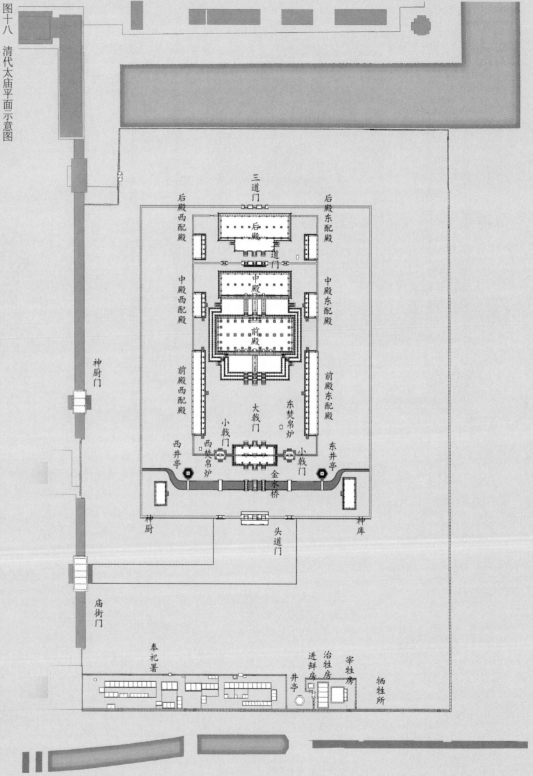

图十八　清代太庙平面示意图

三道门

后殿西配殿　后殿　后殿东配殿

中殿西配殿　中殿　中殿东配殿

前殿西配殿　前殿　前殿东配殿

大戟门

东焚帛炉

小戟门

西焚帛炉　小戟门

西井亭　金水桥　东井亭

神厨门

神厨　头道门　神库

庙街门

奉祀署

治牲房　宰牲房

进鲜房　井亭

牺牲所

Fig.18　Plan sketch of the Ancestral Temple in the Qing Dynasty

Fig.19　*General Map of Altar of Land and Grain in the Collected Statutes of the Great Qing Dynasty* (compiled in the Kangxi Period)

# II Ancestral Temple and Altar of Land and Grain in the Republican Period

## (I) Scope of Altar of Land and Grain and Ancestral Temple after Government of the Republic China took power

After the revolution of 1911, the emperor of the Qing Dynasty abdicated, and according to the stipulations listed in the *Preferential Treatment for Royal Family of the Former Qing Dynasty* the royal family was supposed to move to the Summer Palace. However, the movement was not fulfilled. In 1913, following negotiations with the Qing's royal family, the imperial city area south of the three major halls (excluding the Ancestral Temple) was put under the administration of the Republican government. In autumn of the same year, ZHU Qiqian, the then head of civil administration, proposed to open the Altar of Land and Grain as a park. The park was officially opened on October 10th 1914[45].

The Ancestral Temple remained under the administration of the Qing royal family until the 1924, when the Qing Royal Family Rehabilitation Commission of the Beiyang government took it over. In 1925, the Ancestral Temple came under the jurisdiction of the Palace Museum. In 1926, the Beiyang government renamed the Ancestral Temple "Peace Park" and opened it to the public, and in 1930, the Ancestral Temple was opened to the public as a branch of the Palace Museum. In 1935, the Palace Museum renovated the Ancestral Temple courtyards, built new offices and roads, and erected earthen hills in the eastern district[46] (Fig.20, Fig.21).

## (II) Zhongshan Park and its construction during the Republic of China period

In 1914, the Altar of Land and Grain was opened as the first public garden in the capital and was renamed "Central Park." In the beginning of that year, repairs, renovations and new construction was undertaken in accordance with the functional requirements of the park. The south gate of the park was opened in September, and new roads, hills, pavilions, and office buildings were established in order to transform the temple into a modern park. In September of 1928, it was renamed "Zhongshan Park." In September of 1937, after the Japanese army invaded Beijing, it was renamed "Peking Park," but just one month later, in October, its name was changed back to "Central Park." In 1945, after Beijing was recovered from the Japanese, its name was restored to "Zhongshan Park."

二、民国及之后的太庙与社稷坛

（一）民国政府接管后的社稷坛及太庙区域

辛亥革命后，清帝逊位，根据《清室优待条件》皇室应迁居颐和园，但实际并未履行。民国二年（一九一三年）经由民国政府与清室交涉，将除太庙之外的宫城区域南的三大殿以南的宫城区域划归民国政府管理。同年秋，时任内务总长的朱启钤建议将社稷坛辟为公园，民国三年（一九一四年）十月十日国庆正式开园[45]。

太庙仍由清室管理，直至民国十三年（一九二四年）由北洋政府清室善后委员会接管。民国十四年（一九二五年）太庙归属故宫博物院。民国十五年（一九二六年）北洋政府曾将太庙命名为『和平公园』对外开放。民国十九年（一九三〇年），太庙作为故宫博物院分院对外开放。民国二十四年（一九三五年）前后，故宫博物院对太庙庭院进行了整理、维修，增建办公用房，修筑道路，建造藤萝架，在东区堆垫土山[46]（图二十、图二十一）。

（二）民国时期的中山公园及其建设

民国三年（一九一四年），社稷坛作为京都市内首个公共园林开放，定名为『中央公园』。是年起，根据公园功能要求，开始一系列修缮、整理和新的建设。九月辟建公园南大门，并新建道路、土山、亭室、办公用房，近代公园格局开始呈现。民国七年（一九二八年）九月，奉令改名为『中山公园』。

By the early years of the Republican period, despite being rundown due to many years of neglect, the Altar of Land and Grain maintained its original Ming dynasty layout and structure. After the establishment of the Central Park, more garden buildings and attractions were erected outside the altar and some buildings and walls were also renovated or removed. After the public garden took shape in the late 1910s (Fig.22), except for sporadic renovations and the construction of the "Beijing Music Hall" on the east of the Altar of Land and Grain during the Japanese occupation, the layout of Zhongshan Park remained unchanged until 1949. The Worshiping Hall of the Altar of Land and Grain was converted into the "Zhongshan Memorial Hall" in 1929 because this was where SUN Yat-sen's coffin was kept. The Ji Gate was changed to a hall in 1916 and served as a library. Some aspects of the building were modified, but the wooden structure and roof remained largely intact.[47]

# (III) Ancestral Temple and Altar of Land and Grain in the People's Republic of China

In January 1950, the Government Administration Council of the Central People's Government decided to transfer the administration of the Ancestral Temple to the city of Beijing. On April 10th, the Palace Museum officially transferred the Ancestral Temple to the Beijing Federation of Trade Unions and opened it to the public as the Beijing Working People's Cultural Palace. The architecture and interior decoration was renovated during this time (Fig.23).[48]

After the Communist takeover of Beijing in 1949, the Beijing Military Control Commission took control of Zhongshan Park, and in May 1950 the park was placed under the control of the newly established Beijing Garden Management Committee. Although Zhongshan Park (including the area inside the inner walls) underwent a lot of new construction and reconstruction during this time[49] (Figure 24), the main historical buildings from the Republic of China were kept.

In January 1988, the State Council announced the third batch of key cultural relics protection units under national protection, and the Beijing Ancestral Temple and the Altar of Land and Grain were both listed. In 1987, the Forbidden City of Beijing was included in the World Heritage List, but the "temple in east and altar in west" were not included. Research on the Ancestral Temple and Altar of Land and Grain in the Ming and Qing dynasties was scarce. After ZHANG Guorui's *Ancestral Temple Study* in 1932[50] and the survey and mapping of Beijing's central axis buildings in the early 1940s,[51] no research was carried out on the Ancestral Temple until the 1980s. The gap in research on the Altar of Land and Grain was not filled in 2002, when the Beijing Zhongshan Park Administration published the *Chronicle of Beijing Zhongshan Park*. The lack of related research has negatively impacted our understanding of the cultural heritage value of the "temple in east and altar in west."

## （三）中华人民共和国时期的太庙与社稷坛

一九五〇年一月，中央人民政府政务院决定将太庙拨给北京市，四月十日，故宫博物院正式将太庙移交给北京市总工会，作为北京市劳动人民文化宫对社会开放。为满足文化宫的功能需要，其后陆续对部分文物建筑、其内部装修及太庙墙垣进行了一定程度的改造[48]（图二十三）。

一九四九年北平和平解放之后，由北平市军事管制委员会接管中山公园，一九五〇年五月归属于新成立的北京市园林管理委员会。随着对公园功能要求的提高，特别是改革开放之后，中山公园（包括内垣之内）进行了大量新的建设与改造[49]（图二十四），但主要历史建筑仍维持民国时期修缮或改建原貌。

民国二十六年（一九三七年）日军占领北平后，于九月改称『北平公园』，同年十月恢复『中央公园』原名。民国三十四年（一九四五年）北平光复后，于九月恢复『中山公园』名称。

直至民初年，社稷坛虽因年久失修有破败之相，但依旧维持明清两代原有形制。中央公园成立之后，从民国三年（一九一四年）起，在坛垣之外开始增加了较多的园林建筑与景点，个别失修建筑与垣墙也有所拆除与改动。一九一〇年代末期公共园林格局形成后（图二十二），除在坛垣中零星建设及日据时期于社稷坛东侧建北京市音乐堂外，直至一九四九年中山公园格局基本保持稳定。核心历史建筑中，社稷拜殿因停放过孙中山先生灵柩而于民国二十八年（一九二九年）改建为中山纪念堂，建筑主体则未有改建。戟门则于民国五年（一九一六年）改为殿堂，作为图书馆使用，形制修改较大，但大木结构及屋顶未变。[47]

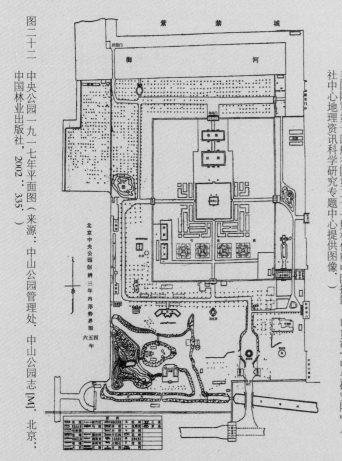

图二十二 中央公园一九一七年平面图（来源：中山公园管理处. 中山公园志[M]. 北京：中国林业出版社，2002：335.）

图二十一 一九四五年日据时期末的紫禁城及「左祖右社」区域（来源：一九四五年一月美国陆军第十四航空队第二十一照相侦察中队摄影，由台湾『中央研究院』人社中心地理资讯科学研究专题中心提供图像。）

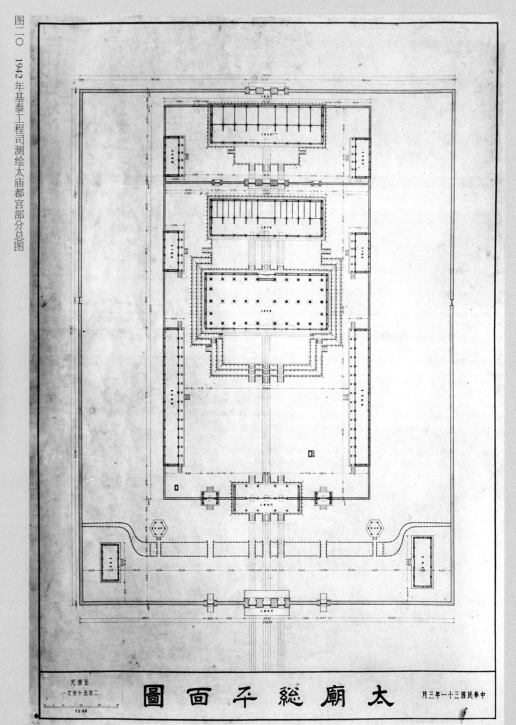

图二〇 1942年基泰工程司测绘太庙都官部分总图

Fig.20　*General Drawing of Palaces of the Ancestral Temple* mapped and surveyed by Kwan, Chu and Yang Architects in 1942

Fig.21　The Forbidden City and the "temple in east and altar in west" area at the end of the Japanese occupation in 1945 (Source: photographed by the 21st Photographic Reconnaissance Squadron of the 14ᵗʰ Air Force of the US Army in January 1945. Image courtesy of GIS Center at Academia Sinica in Taiwan, China.)

Fig.22　Plan of the Central Park in 1917 (Source: Beijing Zhongshan Park Administration office. Chronicle of Beijing Zhongshan Park: Beijing. China Forestry Press, 2002. pp. 335)

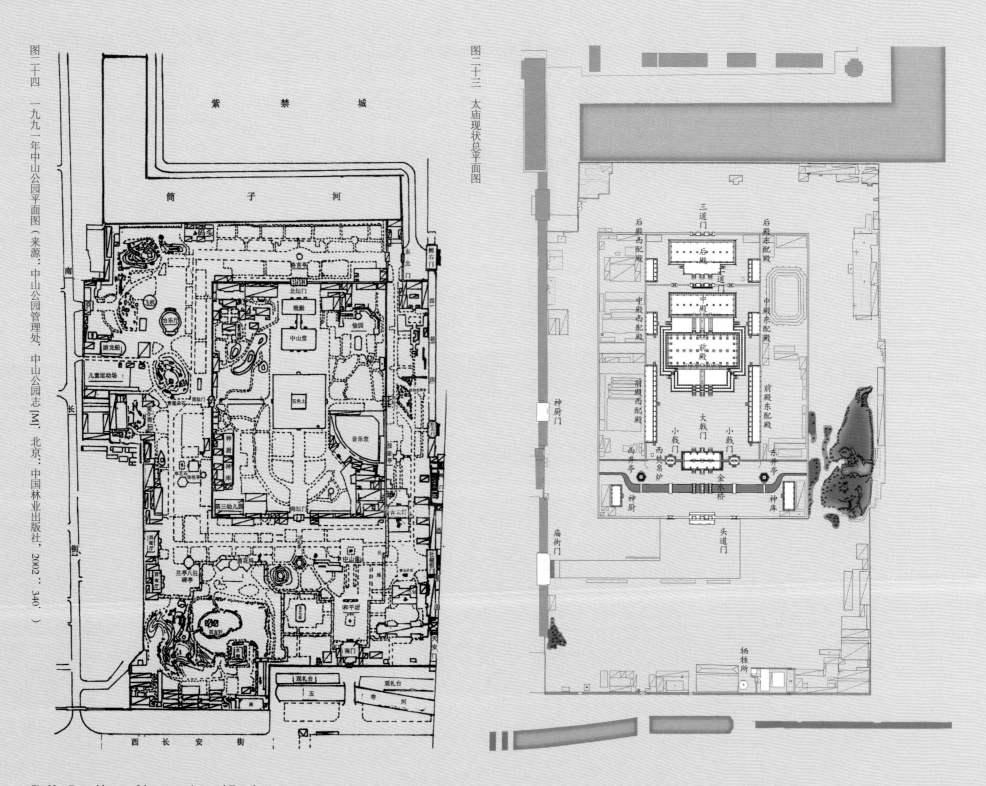

图二十四　一九九一年中山公园平面图（来源：中山公园管理处 中山公园志[M]．北京：中国林业出版社，2002：340．）

图二十三　太庙现状总平面图

Fig.23　General layout of the current Ancestral Temple
Fig.24　Plan of the Central Park in 1991 (Source: Beijing Zhongshan Park Administration office. Chronicle of Beijing Zhongshan Park. Beijing: China Forestry Press, 2002. PP. 340)

From 1996 to 1998, the School of Architecture of Tianjin University carried out systematic surveying and mapping on the historical buildings of the Altar of Land and Grain and the Ancestral Temple, and initiated in-depth research from the beginning of the 20th century.[59] In August 2014, the *Protection Planning of Beijing Ancestral Temple*, which was developed by the Architectural Design and Research Institute of Tianjin University and supported by Beijing Working People's Cultural Palace, was completed.[60]

# III Overview of Cultural Relics and Buildings in Ancestral Temple and Altar of Land and Grain

## (I) Overall layout of the Ancestral Temple

The Beijing Ancestral Temple is located in the southeast of the Forbidden City and on the east side of the imperial path between the Meridian Gate and Tian'anmen, facing the Altar of Land and Grain. Along its north-south central axis, the Ancestral Temple has three gates, the "Quezuo Gate" (literally West Gate of Imperial Palace) opposite to the Meridian Gate, the "Miaoyou Gate" (literally the East Gate of the Temple, also called the Gate of Divine Kitchen) to the north of the Duanmen (the Gate of Uprightness), and the "Taimiaojie Gate" (literally the Street Gate of the Ancestral Temple). All the three gates are oriented to the east-west direction. Quezuo Gate is the main point of entry and exit, while the Taimiaojie Gate was the main entrance for sacrificial rites from the Forbidden City to the Ancestral Temple.

The Ancestral Temple architectural ensemble covers an area of around 16.9 hectares. Facing south, it takes on a rectangular shape, with the main buildings alinged along the central axis from south to north. The buildings in the Ancestral Temple can be divided into three courtyards enclosed by red walls and are capped with yellow-glazed ceramic roof tiles. Three gates with glazed tile roofs are located on the south outer wall of the first courtyard, with one side gate on each side. The courtyard outside the wall is shaded with luxuriant cypress trees, creating a dignified and solemn atmosphere (Fig.25, Fig.26). In the southeast area of the courtyard are the Butchering Pavilions and five Sacrifice Preparation Rooms, all facing west and surrounded by a wall with a west-facing gate. A hexagonal well pavilion sits outside the gate. There used to be a three-bayeast-facing Enshrining House in the southwest area of the courtyard, but it was destroyed. The courtyard still contains three east side rooms and three west side rooms surrounded by a wall with a north-facing gate.

# 三、太庙与社稷坛文物建筑概述

## （一）太庙总体格局

北京太庙位于紫禁城外的东南方、午门至天安门间御道的东侧，隔御道和社稷坛遥遥相对。与御道连接从北至南分设太庙西侧午门前的『阙左门』、端门北的『庙右门』（又称『神厨门』），以及天安门和端门之间御路东侧的『太庙街门』，三门均东西向布局。阙左门为通常出入之路径，太庙街门由紫禁城前往太庙进行祭祀的行进路线之正门。

一九八八年一月，国务院颁布第三批全国重点文物保护单位名录，北京太庙和社稷坛同时在列。

一九八七年北京故宫入选世界遗产名录，但『左祖右社』并不在登录范围之内。对于明清太庙和社稷坛文物建筑的研究较为稀缺，在一九三二年张国瑞《太庙考略》和一九四〇年代初期的北京中轴线建筑测绘之后，直至一九八〇年代才再次出现系统化的太庙研究[61]，而社稷坛的系统研究直到二〇〇二年中山公园管理处出版的《中山公园志》方对空白有所填补。相关研究的缺失，对于『左祖右社』的文化遗产价值的理解和认定产生了很大的负面影响。

天津大学建筑学院自一九九六年至一九九八年对社稷坛和太庙历史建筑进行系统的测绘工作，并基于此项测绘工作于二十一世纪初开始进行了一系列深化研究[62]。二〇一四年八月，由天津大学建筑设计研究院编制，北京市劳动人民文化宫参编的《北京太庙保护规划》完成[63]。

The main buildings of the Ancestral Temple are located in the second courtyard, which has the Daji and Xiaoji gates in the southern wall. To the south of the Daji Gate (literally the Gate of Big Halberd) are 5 single-arch white stone bridges. On each of the east and west sides to the north of the bridges is a hexagonal well pavilion. A Divine Store is located on the east side to the south of the bridges, and a Divine Kitchen is located on the west side. Behind the Daji Gate is the eleven-bay Front Hall (the Sacrificial Hall), which is capped with a double-eave hip roof. With a three-layered *xumizuo* base, the hall has a spacious front platform and its rear base connects with that of the Central Hall. To better highlight the solemn function of the temple, all the beams and columns in the Front Hall are made from whole pieces of Phoebe sheareri. The roofs, ceilings and four pillars of the central and flanking bays are all decorated with gold designs, without any colored paintings. In this hall the emperor offered sacrifices to the ancestors during grand rituals. There are fifteen single-eave hip-and-gable-roofed side halls on the left and right wings of the Front Hall. Memorial tablets of princes and aristocrats were enshrined in the east side halls, while meritorious statesmen were enshrined in the west side halls. The Central Hall (the Bedroom Hall) has 9 bays, with a single-eave hip roof. Memorial tablets of deceased emperors and empresses were enshrined here. Five single-eave hip-and-gable-roofed side halls, where sacrificial utensils were stored, line each wing. The shape and structure of the Rear Hall is similar to that of the Central Hall. As a temple to worship remote ancestors, the Rear Hall is separated from the Central Hall by a wall. Three gates with glazed tile roofs are located in the middle of the wall, with one gate on each side. There are also three gates in the middle of the back wall of the Rear Hall.

On the day of the sacrifice, the imperial procession guard lined up outside the Meridian Gate to await the emperor's ritual carriage (*fajia*). About one hour before the sunrise, the emperor would depart from the Gate of Heavenly Purity for the Gate of Supreme Harmony, where the emperor would alight and transfer to the golden imperial carriage. Drums were beaten at the Meridian Gate to declare martial law, and the honor guard led the way. Princes and officials of lower levels who did not accompany the emperor for the sacrifice knelt down to see off the procession. The emperor entered through the left Taimiaojie Gate and got off on the right side of the Scared Way outside the south gate of the Ancestral Temple. The master of ceremony handled the ceremonial procession and led the emperor to enter through the east gate of the Ancestral Temple to the east room outside Jimen Gate to worship the memorial tablets there. Then the master of ceremony would petition the emperor to curtsey, after which the emperor would exit the room. After a ritual hand washing, the master of ceremony would lead the emperor through the east gate of the Jimen Gate to ascend the east staircase of the Ancestral Temple and arrive at the memorial tablets through the east door of the Front Hall, with the emperor standing facing the north.[54]

太庙建筑群占地面积约十六点九公顷，坐北朝南，平面呈长方形，主要建筑由南向北依次排列于中轴线上。太庙建筑群被三道黄琉璃瓦顶的红色墙垣分隔成三个封闭式的院落。第一层院落的外垣正南辟琉璃花门三道，左右旁门各一道。垣外院落植满柏树，浓荫蔽日，枝繁叶茂，营造出一种凝重、庄严的气氛（图二十五、图二十六）。院落东南有宰牲亭三间，治牲房五间，均西向，垣一重，门一，西向。门外有六角井亭一。西南有奉祀署，面阔三间，东向，现已不存。左右房各三间，垣一重，门一，北向。

太庙的主要建筑集中于第二进院落中，墙垣南侧群有大小戟门。大戟门南侧有单孔白石拱桥五座，桥北面东西两侧各有一座六角井亭，桥南东侧为神库，西侧为神厨。大戟门后为前殿（享殿），面阔十一间，重檐庑殿顶。台基须弥座三重，殿前月台宽阔，殿后台基与中殿台基相连。为更好地突出宗庙祭祀性建筑的特色与效果，前殿梁柱均为整料金丝楠木。明间和次间的殿顶、天花、四柱全部贴赤金花，不用彩画装饰，有意避开浓艳华丽的暖调，而以清淡雅致的冷调代替，凸显出宗庙祭祀的特殊氛围。每次大祭时皇帝在此祭祀先祖列帝。前殿左右两厢有配殿各十五间，单檐歇山顶，东配殿祀配享王公，西配殿祀配享功臣。中殿（寝殿）九间，单檐庑殿顶，殿内供奉列圣、列后神龛。左右配殿各五间，单檐歇山顶，用以存贮祭器。后殿形制与中殿相似，作为祭祀远祖神主的祧庙，与中殿之间有墙垣相隔。墙正中辟琉璃花门三道，左右旁门各一。后殿后围墙正中辟琉璃花门三道。

祭祀之日，銮仪卫陈设法驾卤簿于午门外，设金辇于太和门阶下。日出前四刻，皇帝御祭服乘礼舆自乾清门出宫，至太和门外，降礼舆换乘金辇。午门严鼓，法驾卤簿前导，不陪祀的王以下各官齐集朝服跪送。皇帝入太庙街门左门，至太庙南门外降舆，赞引太常卿一人引导皇帝由太庙南门左门入，至戟门外东间幄次神主奉安完毕，太常寺卿奏请行礼，皇帝出幄次，盥洗，赞引太常卿引导皇帝由戟门左门入，升太庙左阶，入前殿左门，就拜位前，北向立。[五十四]

图二十六 太庙历史航片（来源：北京市劳动人民文化宫提供）

图二十五 国家图书馆藏样式雷图《太庙全图画样》中所反映的树木栽植意向

Fig.25　The plantation plan shown in Design Studio Lei's *Full Drawing of the Ancestral Temple* collected in the National Library of China

Fig.26　Historical aerophotograph of the Ancestral Temple (Source: Provided by Beijing Working Pepole's Cultural Palace )

## (II) Main buildings on the central axis of the Ancestral Temple

The Daji Gate of the Ancestral Temple is five bays in width and two bays in depth. It is a nine-purlin palace-like structure, covered a single-eave hip roof covered by yellow glazed ceramic tiles (Fig.27). The roof beams concealed above the ceiling were of a standard post and lintel construction, with the third, fifth, seventh and ninth *nanmu* girders trussed in order, showing obvious inclination. Aspects of its wooden structure and bracket sets, as well as existing architectural paintings carry distinct characteristics of the Ming Dynasty. The Xiaoji Gates on both sides follow the central-parting structure with a single space each, covered by single-eave hip-and-gable roof. From the characteristics of incline and the tapering peristyle columns, as well as specific characteristics of the bracket sets, sparrow braces, and architectural paintings, we estimate which is an original Ming Dynasty structure.

The Front Hall (Sacrificial Hall) of the Ancestral Temple follows a central-parting in-plane structure with three columns. It has nine spaces in width and four spaces in depth, coupled with an ambulatory. The hall takes on a large palace-like shape with eleven purlins, yellow-colored glaze tiles, and a hip roof with double eaves (Fig.28). In the heavy timber structure of the Front Hall, the upper roof eave and the column frame layer are separated, each as a self-contained whole structure, mirroring the characteristics of the hall structure in the Ming Dynasty.[55] The depths of the corridor, lower purlin tie-beam, middle purlin tie-beam, upper purlin tie-beam and ridge tie-beams of the Front Hall of the Ancestral Temple vary, unlike the we see in Song and Qing period buildings in which the distance between the steps of the purlins is the same . Variation in sizes of the steps between the purlins is a common feature of official architecture in the Ming Dynasty. From the pitch and curvature projection of the roofing, we can see that the line is comparatively smooth, which is similar to that in the Song Dynasty but different from that in the Qing Dynasty when the curved angles were at ratios of integers to present obvious break points.

Meanwhile, the Front Hall's cap-block mortise openings are of around 12.5cm long, or 4.0 cun (a unit of length in ancient Chinese architecture), being the largest among the existing buildings from the Ming and Qing dynasties. The method of constructing the upper eave bracket sets was also a commonly used approach during the Ming Dynasty. Except for the heavy timber structure, colored paintings on the ridge purlins, and purlin-attached ties in the central bay of the concealed roof beams of the Front Hall were a common practice in the hip-roof palaces and halls of very high grades in the Ming Dynasty. The Daji Gate, Front Hall, Central Hall and Rear Hall of the Ancestral Temple all show this characteristic. Specifically, the paintings on the ridge purlins and purlin-attached tie-beams in the central space of the Front Hall are of the highest level, and the painting on the ridge purlins in the central and side spaces of the roof beams all reflect the Ming style. Finally, from the materials used for the Front Hall of the Ancestral Temple, its lower column frames all adopt whole-piece *nanmu* components to form the main structure. By the Qing dynasty, the construction of such a large-

## （二）太庙中轴线主要建筑

太庙大戟门为五间二进、分心槽九檩门殿式建筑，黄琉璃筒瓦，单檐庑殿顶屋面，庑殿顶不推山（图二十七）。草架为规整的抬梁式构架，三、五、七、九架梁依次叠架，楠木材质，柱有明显侧脚做法。大木构架及斗栱细节、现存彩画等具有明确的明代特征。两侧的小戟门为单开间分心造、单檐歇山顶。从檐柱有侧脚及收分，以及斗栱、雀替、彩画等特点来看，亦应为明代原物。

太庙前殿（享殿）平面为分心槽用三柱，殿身九间四进加副阶周匝。为十一檩大型殿堂形制，黄琉璃筒瓦，重檐庑殿顶屋面，屋顶有推山（图二十八）。前殿的大木结构中，上檐屋盖层与柱框层是上下分开、自成整体构架的，反映了明代殿阁式构架的特征[55]。太庙前殿屋盖层从檐步、下金、中金、上金步直至脊步架深均不相等，与宋、清规定的各步架相等不同，而各步架进深不等的特点在明代官式建筑中又极为普遍。从屋面举折投影看，折线较为圆转，亦表现为与宋式接近而不同于清式之各步举折呈整数比、折点较明显之特征。同时前殿斗口取值约为十二点五厘米，合四营造寸，是现存明清两代官式建筑中取值最大者。上檐斗栱做法也是明代重要殿堂建筑斗栱出跳较多时常用的一种方式。除去大木结构之外，在前殿草架明间脊檩及随檩枋上绘制彩画是明代在一些等级特别高的庑殿顶殿堂建筑中的共同做法，太庙大戟门、前殿、中殿、后殿四座建筑均具备这一特点。其中前殿脊檩及随檩枋上的彩画等级可谓最高，草架中明次间脊檩彩画均反映明式风格。最后，从太庙前殿的用材来看，其下部柱框层均系整料楠木构件组成之巨构，如此规模的单体建筑建设，清初开始已无力达成。且大木构件加工的工艺精致细腻，甚至在人眼平常难及之梁架草架部分，其加工仍十分细致。综合以上特点，太庙前殿可认为是明代原构。

scale building of *nanmu* was impossible due to a shortage of timbers growing close to major waterways in the forests of southwest China. The processing techniques for the heavy timber components were quite exquisite and delicate. Even the roof beams concealed above the ceiling were also finely processed. To sum up, the Front Hall of the Ancestral Temple can be deemed as an original structure from the Ming Dynasty.

The Central Hall (Bedroom Hall) and the Rear Hall (Remote Ancestor Temple) share basically the same structure. Both are nine bays in width and four bays in depth. Their roofs comprise eleven purlins and have single-eave hip roofs covered by yellow-colored glaze tiles (Fig.29). The Central Hall and the Rear Hall are similar to the Front Hall in terms of using whole-piece nanmu as the construction material, concealed roof beams and a visible truss structured in a clear hierarchy. The characteristics of the steps of the trusses, the 4.0 *cun* cap block mortise openings, the cap block mortise making techniques, and the style of the architectural painting, are all analogous to the Front Hall. Therefore, the two rear halls are also deemed as original Ming Dynasty structures.

# (III) Overall layout of the Altar of Land and Grain

The Altar of Land and Grain is located to the west of the Queyou Gate on the east side of the Meridian Gate. Apart from the Queyou Gate, the altar is connected to the Forbidden City through the Shezuo Gate (the gate on the east of the altar) on the south of the Duanmen (the Gate of Uprightness) as well as the Shejijie Gate (the street gate of the Altar of Land and Grain) between the Tian'anmen and the Duanmen (the Gate of Uprightness) and on the west side of the imperial path. All the three gates face west.

The Altar of Land and Grain faces north and takes on a rectangular shape. The God of Land and the God of Grain share the same parapet and are surrounded by three walls, namely the parapet, the altar wall and the outer wall. The Five Colors Earth Altar is located in the center, situated on the central axis with the Worshiping Hall (the Sacrificial Hall in the Ming Dynasty), and the Jimen (the Worshiping Hall in the Ming Dynasty) aligned from south to north. The upper side length is 15.95meters, the middle side length is 16.90meters, and the lower side is 17.85 meters. A green-and-white marble staircase is built in the middle of each of the four sides and contain four stairs each. The platform is paved with tiles of five colors to signify the five directions: green in the east, red in the south, white in the west, black in the north and yellow in the middle. The use of colored tiles on the parapet also corresponds to the cardinal direction. In the Ming Dynasty, the parapet was employed colored bricks, but

中殿（寝殿）及后殿（祧庙）构架基本相同，均为九间四进、分心槽十一檩殿堂建筑。黄琉璃筒瓦，单檐庑殿顶，屋顶有推山（图二十九）。与前殿相较，两殿除去单檐与重檐的区别之外，从整料楠木用材、草架与明架分层明确、步架与举架特征、斗口尺寸为四营造寸、斗栱做法，以及彩画形式等均同前殿相似。两殿也应为明代原构。

（三）社稷坛总体格局

社稷坛建于午门东侧『阙右门』以西，与紫禁城的联系尚有端门南『社左门』及天安门和端门之间御路西侧的社稷街门，三门均西向。

社稷坛建筑群北向布局，平面呈长方形，社、稷同坛同壝，有壝垣、坛垣、外垣三重。五色土坛居中，拜殿（明代祭殿）、戟门（明代拜殿）往北依次排列在中轴线上。坛方形，现存三层，上层边长十五点九五米，中层边长十六点九零米，下层边长十七点八五米。四面各设一部垂带台阶，居中，四步，青白石砌筑。台上随方布色：东青、南红、西白、北黑、中黄，铺设五色土。墙墙随四方色，明代以砖砌涂色，清乾隆二十一年（一七五六年）改为四色琉璃砖瓦。墙垣四面居中各设一棂星门，每门原有朱扉两扇（五十八）。

图二十七　大戟门历史照片（来源：北京市劳动人民文化宫提供）

图二十八　从大戟门明间望前殿

Fig.27　Historical photo of Daji Gate (Source: Provided by Beijing Working Pepole's Cultural Palace)
Fig.28　A glimpse of the Front Hall from the central bay of the Daji Gate
Fig.29　Rear Hall of the Ancestral Temple

in 1756, the parapet was changed to glazed bricks covered with colored tiles. Each of the four sides of the parapet has a gate, which used to be equipped with red doors. [56]

Although there is no vegetation inside the altar walls, many cypress trees are planted outside in neat rows. In the southwest corner inside the altar are five Divine Kitchens and five Divine Stores side-by-side facing the east. Divine Kitchens are in the north and Divine Stores are in the south, all covered by overhanging gable roofs with yellow glazed tiles. The Butchering Pavilion is located to the south of the Xitan Gate, with its north, west and south sides enclosed by walls and its east side connected to the altar wall. The rectalinear pavilion has a double-eave gable-and-hip roof of yellow glazed tiles. Overall, the Altar of Land and Grain buildings are plain and elegant. The enclosed courtyards and thick ancient cypress trees create a solemn and dignified atmosphere, which is consistent with its purpose for imperial sacrifices.

The springs and autumns in the Qing Dynasty were the times when the emperor would personally attend the sacrificial rites. The master of ceremony reported the time one hour before the sunrise, and the emperor would take the ritual carriage from the Gate of Heavenly Purity to the Gate of Supreme Harmony, where he would transfer to the golden imperial carriage. Bells were rung at the Meridian Gate to declare martial law, and the honor guard led the way. Princes and officials of lower levels who did not accompany the emperor on the sacrifice knelt down wearing court dress to see off the procession. The golden imperial carriage exited through the Meridian Gate and the Queyou Gate and was then set out of the Beitan Gate (the north gate of the altar). The procession entered through the right gate of the Beitan Gate (the north gate of the parapet) and waited in the Ji Gate. After the rites for the memorial tablets were over, the master of ceremony would instruct the emperor to curtsey, after which the emperor would wash his hands and leave the Beiwei Gate,. When the emperor failed to be present in person, the sacrificial rite would be attended by a dispatched official. On the day of the rite, the dispatched official entered through the Shejijie Gate and then turned to the west to Nantan Gate (the south gate of the altar) to head westward along the parapet. Two ceremony specialists would lead the dispatched official to the curtseying site outside the Beiwei Gate. [57]

## (IV) Worshiping Hall and Jimen of the Altar of Land and Grain

The Worshiping Hall and the Ji Gate are the main buildings of the Altar of Land and Grain. They function as the places for "curtseying" and "court dressing" for the rituals that take place at the Altar of Land and Grain system of the early Ming Dynasty. [58] The name "Worshiping Hall" was used during the Qing, [59] while "Sacrificial Hall" or "Memorial Hall" were used in the Ming. [60] Ji Gate was also called "Ji Hall" [60] and was the name used in the Qing. In the Ming, it was called the "Worshiping

坛垣内原无植被，垣外遍植柏树，井然森列。坛内西南隅有神厨、神库各五间并列，东向。神厨在北，神库在南，悬山黄琉璃瓦屋面。宰牲亭位于西坛门外以南，北、西、南三面围以矮墙，墙之东端与坛墙相接，方形，重檐，歇山黄琉璃瓦屋面。社稷坛整体建筑古朴典雅，加上封闭的院落，浓郁的古柏，衬托出一种肃穆庄重的氛围，与皇家祭祀建筑的性质相一致。

清代春秋常祀皇帝亲祭祀日，日出前四刻太常卿告时，皇帝御祭服乘礼与由乾清门出宫，至太和门阶下降舆乘金辇。午门鸣钟，法驾卤簿前导，不陪祀的王以下各官齐集朝服跪送。金辇出午门，经阙右门至北坛门外降辇。入北坛门右门，候于戟门幄次。待奉安神位毕，太常寺卿奏请行礼，盥洗后帝至北墙门外拜位，南向立。因事祇告社稷礼则由遣官祭祀，祀日遣官由社稷坛街门入，北拐西转入南坛门后沿墙西行，由赞礼郎二人引导至墙北门外拜位 [55]。

## （四）社稷坛拜殿与戟门

拜殿与戟门是社稷坛主体建筑，明初社稷坛制度为「行礼」「具服」二殿设置 [56]。拜殿为「拜殿」，清代名称 [57]，明代称「享殿」或「祭殿」 [60]。戟门又称「戟殿」 [61]，为清代名称，明代为「拜殿」。两殿设置遵循南京社稷坛制度，但名称上于后世多有混淆 [62]。

Hall."⑥ Both halls were modeled on the Nanjing Altar of Land and Grain system, but their names were confused in later generations.⑥

The current Worshiping Hall was built in the early Ming Dynasty and is located due north of the altar parapet, serving as the place for curtseying during windy and rainy days of sacrifices.⑥ The Memorial Hall is situated atop single story base. The hall is five bays in width and three bays in depth and without a colonnades. It is capped by a single-eave hip-and-gable roof with yellow glazed tiles (Fig.30). With three spaces in depth partitioned by four columns, the hall features sagittal symmetry. The eleven timber purlins constitute the open-timbered hall-type construction with penetrating tie beams, among other components. *Chashou* (inverted V-shaped braces) and brackets were omitted in the roof truss to simplify the *tuofeng* (camel-hump shaped support) designs, while camel-hump shaped or camel-hump patterned imposts and short columns raising up the beams from below. On the eave columns are seven-footing mono-petal dual-lever tailed gilded cap block mortises of primitive simplicity. The eave columns and the hall are connected by exposed tie-beams and glided cap block mortises to strengthen the horizontal stability. The cap block mortises of the corner bracket sets adopt the mandarin-ducks-crossing-necks bracket arms with carved patterns. Under the bracket arms are herringbone flutes imitating the bracket sets, and the smaller bracket ends do not extend to support the upward bracket arms. Instead, they hide behind the upper layer of bracket arms. Such a practice is roughly in line with the construction techniques of the Song Dynasty. The nose and center block structuring methods also show typical characteristics of the early Ming Dynasty. In addition, while the heavy timber structure is simpler, it attaches more importance to the connections between various components. Cap block mortises are placed under purlins, dividing bracket sets are placed between cross beams and beam-attached ties, and glided cap block mortises are used to connect eaves and purlin tie-beams. The exquisite processing of the timber members and the tight connection of components, as well as the detail-oriented carving and decorations, add to the aesthetic value of the interior space while pursuing simplicity and practicability.

Ji Gate is located to the north of the Memorial Hall and served th as the place where the emperor would dress for the sacrifice. It constructed in 1425. Its main wooden construction has survived till today, although it has undergone multiple renovations and repairs. The Memorial Hall is situated on a single story base. The hall is five baysin width and two bays in depth and does not have a colonnade. It is covered by a single-eave hip-and-gable roof of yellow glazed tiles (Fig.31). Its central and side bays are enclosed by doors. The front row columns of the central and side bays are omitted and the beams extend straight to the cornices. The eleven heavy timber purlins form the open-timbered hall-type construction, with the architectural features similar to those of the Memorial Hall. The three

现存拜殿建于明初，位于坛壝正北，为祭祀日遇风雨时行礼之处（图三十四）。祭殿单层台基，面阔五间，进深三间，平面无廊。单檐歇山顶，黄琉璃瓦屋面（图三十）。前后三进四柱，前后对称。十一檩大木为彻上明造的厅堂式构架，采用了随梁枋、穿插枋等构件，屋架上取消了叉手、托脚，简化了驼峰式样，代之以梁栿端头直接承檩、梁下垫以驼峰或隐刻驼峰式样的驼墩、童柱。檐柱上置七踩单翘重昂后尾溜金斗栱，做法古朴。檐柱与殿身檐柱间以桃尖梁及溜金斗栱连接，旨在加强横向构架的稳定性。角科斗栱用鸳鸯绞手栱，刻于栱身，栱下皮则模仿栱身成人字形凹槽，这时的小栱头一般不伸出承托上跳栱身，而是隐于上层栱身之后，做法与宋《营造法式》制度基本相同。加之耍头与齐心斗的做法，具有典型的明初建筑特色。另外，在大梁与随梁枋间设隔架科，以溜金斗栱联系檐、金步等。由于大木加工细致，构件结合紧密，加之注重装饰细节的刻画，因此大木构架在追求简洁实用的同时，也增加了室内空间的艺术欣赏价值。

戟门位于祭殿北，具服之所。建成于明洪熙元年（一四二五年），后经多次修缮，现存大木构架仍系明代原物。单层台基，面阔五间，进深二间，平面无廊。单檐歇山顶，黄琉璃瓦屋面（图三十一）。

doors in the middle of the Ji Gate were removed in 1916 and sill walls were built along the front and rear eaves. Glass doors and windows were installed. During this time Ji Gate was used as a library. In 1950, the library was moved out of Zhongshan Park. The building now serves as the office of the Beijing municipal committee of CPPCC.[69]

# IV. Surveying and mapping of buildings in the Ancestral Temple and the Altar of Land and Grain

In the early 1940s, ZHANG Bo, an architect with Kwan, Chu and Yang Architects, organized the teachers and students from the Department of Architecture of the Tianjin Institute of Commerce and Technology and the Construction Department under the College of Engineering of Peking University, as well as some employees of Kwan, Chu and Yang Architects to conduct a field surveying and mapping of buildings along the central axis of the Imperial Palace and its peripheral buildings including, the Ancestral Temple and Altar of Land and Grain (Fig.32). In the late 1980s, QI Yingtao with the China National Institute of Cultural Property (CNICP) surveyed and mapped the four buildings along the central axis of the Ancestral Temple as well as the Altar of Land and Grain, the Worshiping Hall, and the Ji Gate during his investigation of Beijing architecture of the Ming Dynasty[69].

At the end of the last century, to deepen the research on Beijing's "temple in east and altar in west" system and meet the cultural heritage protection requirements for cultural relics and buildings, namely to "have protection scope and construction control zone, have signs and descriptions, have records and files, and have dedicated agencies or personnel," Tianjin University, Beijing Working People's Cultural Palace, and Zhongshan Park Administration worked together to survey and map all the historical buildings of the Ancestral Temple and the Altar of Land and Grain in a systematic manner. Specifically, all the historical buildings and some modern buildings were surveyed in 1996, with more than 130 drawings made. From 1997 to 1998, the surveying and mapping of all the historical buildings in the Ancestral Temple were completed, with more than 200 drawings made. Later, institutions such as the Beijing Institute of Ancient Architecture carried out surveying and mapping of a few historic buildings in the Ancestral Temple, supplementing the cultural relics and construction materials.

During its surveying and mapping of the Ancestral Temple and the Altar of Land and Grain, Tianjin University combined traditional surveying and mapping methods with advanced computer mapping methods. In 1996, the surveying and mapping results of the Five Colors Earth Altar were recorded

# 四、太庙与社稷坛的文物建筑测绘

一九四〇年代初期，基泰工程司建筑师张镈组织天津工商学院建筑系和北京大学工学院建工系师生及基泰公司部分员工，对包括太庙和社稷坛在内的故宫中轴线及其外围文物建筑展开了实地测绘（图三十二）。一九八〇年代后期，中国文物研究所的祁英涛在对北京明代建筑进行调查时，对太庙中轴线上的四座建筑及社稷坛拜殿、戟门进行了测绘工作[64]。

二十世纪末期，出于对北京宫城「左祖右社」的深化研究与文物建筑「四有」等文化遗产保护的需求，天津大学与北京市劳动人民文化宫及中山公园管理处合作，对太庙和社稷坛全部历史建筑进行了系统测绘。其中一九九六年进行了所有历史建筑及部分近现代建筑的测绘，绘制图纸二百余幅。一九九七年至一九九八年完成了太庙所有历史建筑的测绘，绘制图纸一百三十余幅。此后尚有北京市古代建筑研究所等机构进行了太庙少数历史建筑的测绘工作，补充了文物建筑资料。

明次间为大门，明间及次间构架省去前金柱，梁栿直接伸至檐口。九檩大木为彻上明造的厅堂式构架，大木建筑特征与祭殿相似。戟门于民国五年（一九一六年）撤去中间三间大门，前后檐砌槛墙，安装玻璃门窗，改作图书馆。一九五〇年图书馆迁出中山公园，现为北京市政协所在地[65]。

with a three-dimensional model. In 1998, all the drawings for the Ancestral Temple were made digitally, the first of its kind in the cultural relics world to adopt computer-aided graphics,. The digital approach provided a new way of archival construction of ancient buildings and a new mode of data collection for cultural relics protection. The surveying and mapping results also provided a solid foundation for Beijing's "temple in east and altar in west" research and subsequent cultural heritage protection.

The book includes around 150 representative drawings selected from the surveying and mapping results, covering all the historic timber buildings of the Ancestral Temple and other structures, such as the Golden Water Bridge and the Silk Burner, as well as ancillary buildings such as the Ji Gate of the Altar of Land and Grain, the Worshiping Hall, and the altar gates. The architectural drawings are arranged in the spatial sequence of the functions of the Ancestral Temple and the Altar of Land and Grain. The drawing system of various core historical buildings of "temple in east and altar in west" is complete and can present a general view of the historical buildings of the Ancestral Temple and the Altar of Land and Grain.

天津大学的太庙与社稷坛测绘过程中结合了传统测绘方式与先进的计算机制图方法：一九九六年五色土坛以三维模型记录测绘成果；一九九八年太庙的所有图纸均以计算机绘制，是文物测绘领域中系统使用计算机绘图最早的范例之一，也提供了新的古建筑档案建设方式及文物保护数据采集模式。

这一批测绘成果也为北京『左祖右社』的研究，以及后续的文化遗产保护工作提供了坚实的基础。

本书所收录的图纸从这一批测绘成果中精选约一百五十幅代表图纸，涵盖太庙所有古建筑及金水桥、焚帛炉等设施，以及社稷坛戟门、拜殿和坛门等附属建筑。依太庙和社稷坛建筑组群使用功能的空间顺序编排各个建筑的图纸。『左祖右社』各核心历史建筑的图纸系统完整，能够充分体现太庙与社稷坛的历史建筑全貌。

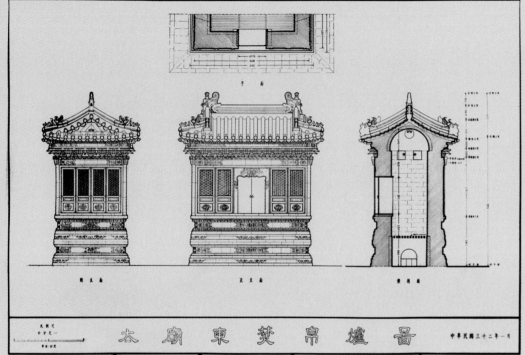

图三十二　民国三十二年（一九四三年）测绘的太庙东焚帛炉图纸

太 庙 東 焚 帛 爐 圖

图三十　社稷坛拜殿

图三十一　社稷坛戟门现状

Fig.30　Worshiping Hall of Altar of Land and Grain
Fig.31　Today's Jimen of the Altar of Land and Grain
Fig.32　Drawing of the Eastern Silk Burner of the Ancestral Temple surveyed and mapped in the 32nd year of the Republic of China Period (1943)

# 注 释

（一）如：『小宗伯之职，掌建国之神位。右社稷，左宗庙』（周礼·春官·宗伯）；『匠人营国……左祖右社』（周礼·冬官·考工记）；『建国之神位，右社稷，而左宗庙』（礼记·祭义）；等。

（二）姜波·汉唐都城礼制建筑研究[M]·北京：文物出版社，2003·

（三）明太祖实录，卷21·

（四）明太祖实录，卷24·

（五）大明集礼，卷8·

（六）明太祖实录，卷25·

（七）明史·礼制，卷51·

（八）明史·礼制，卷100·

（九）明史·礼制，卷51·

（十）明太祖实录，卷60·

（十一）明太祖实录，卷51·

（十二）明史·礼制，卷114·

（十三）明史·礼制，卷51·

（十四）明一统志，卷1·

（十五）明一统志，卷1·

（十六）春明梦余录，卷17·

（十七）曹鹏·明代都城坛庙建筑研究[D]·天津：天津大学博士学位论文，2011·

（十八）明英宗实录，卷135·

（十九）明英宗实录，卷277·

（二十）明孝宗实录，卷1·

（二十一）明孝宗实录，卷7·

（二十二）明武宗实录，卷185·

（二十三）中山公园管理处·中山公园志[M]·北京：中国林业出版社，2002·

（二十四）明武宗实录，卷109·

（二十五）明世宗实录，卷39·

（二十六）明世宗实录，卷51·

（二十七）明世宗实录，卷74·

（二十八）明世宗实录，卷130·

（二十九）明世宗实录，卷166·

（三十）明世宗实录，卷167·

（三十一）明世宗实录，卷171·

（三十二）明世宗实录，卷191、192、193·

（三十三）明会典，卷89·

（三十四）曹鹏·明代都城坛庙建筑研究[D]·天津：天津大学博士学位论文，2011·

（三十五）明世宗实录，卷216·

（三十六）明世宗实录，卷248·

（三十七）明世宗实录，卷280·

（三十八）明世宗实录，卷300·

（三十九）曹鹏·明代都城坛庙建筑研究[D]·天津：天津大学博士学位论文，2011·

（四十）清世祖实录，卷8·

（四十一）清朝通典，卷45·

（四十二）中国第一历史档案馆·奏销档，267册·

（四十三）闫凯·北京太庙建筑研究[D]·天津：天津大学硕士学位论文，2004·

（四十四）大清会典事例，卷427·

（四十四）中山公园管理处. 中山公园志 [M]. 北京：中国林业出版社，2002.

（四十五）中山公园管理处. 中山公园志 [M]. 北京：中国林业出版社，2002.

（四十六）天津大学建筑设计研究院、北京市劳动人民文化宫. 全国重点文物保护单位北京太庙保护规划，2014.

（四十七）中山公园管理处. 中山公园志 [M]. 北京：中国林业出版社，2002.

（四十八）天津大学建筑设计研究院、北京市劳动人民文化宫. 全国重点文物保护单位北京太庙保护规划，2014.

（四十九）天津大学建筑设计研究院、北京市劳动人民文化宫. 全国重点文物保护单位北京太庙保护规划，2014.

（五十）中山公园管理处. 中山公园志 [M]. 北京：中国林业出版社，2002.

（五十一）张国瑞. 太庙考略 [M]. 北京：故宫博物院，1932.

（五十二）参见：姜舜源. 清代的宗庙制度 [J]. 故宫博物院院刊，1987(3)：15-23+57；傅公钺. 清代的太庙 [J]. 故宫博物院院刊，1986(3)：73-79+86.

（五十三）参见：闫凯. 北京太庙建筑研究 [D]. 天津：天津大学硕士学位论文，2004；亚白杨. 北京社稷坛建筑研究 [D]. 天津：天津大学硕士学位论文，2005；池小燕，曹鹏. 北京地坛与社稷坛祭祀对象之辨——略述地祇神与社稷神各历史时期发展演变 [J]. 沈阳建筑大学学报（社会科学版），2006(4)：309-312；闫凯，王其亨，曹鹏. 北京明清皇家三大殿之比较研究 [J]. 山东建筑工程学院学报，2006(2)：116-121；亚白杨. 明清社稷坛空间布局设计意向——「左祖右社」格局探源 [J]. 古建园林技术，2010(2)：57-59；曹鹏. 明代都城坛庙建筑研究 [D]. 天津：天津大学博士学位论文，2011；等.

（五十四）天津大学建筑设计研究院、北京市劳动人民文化宫. 全国重点文物保护单位北京太庙保护规划，2014.

（五十五）钦定大清会典，卷40.

（五十六）郭华瑜. 明代官式建筑大木作研究 [M]. 南京：东南大学出版社，2005.

（五十七）钦定大清会典，卷71.

（五十八）钦定大清会典，卷43.

（五十九）明一统志，卷1.

（六十）钦定大清会典，卷71.

（六十一）大明会典，卷85.

（六十二）高宗纯皇帝实录，卷921.

（六十三）钦定大清会典，卷71.

（六十四）曹鹏. 明代都城坛庙建筑研究 [D]. 天津：天津大学博士学位论文，2011.

（六十五）明太祖实录，卷114.

（六十六）中山公园管理处. 中山公园志 [M]. 北京：中国林业出版社，2002.

祁英涛. 北京明代殿式木结构建筑构架形制初探 [M]//祁英涛. 祁英涛古建论文集. 北京：华夏出版社，1992：325-341.

# Notes

① For example, "Xiao Zongbo position is in charge of building memorial tablets of the state. Altar of Land and Grain is in the west, and Ancestral Temple is in the east." (Rites of Zhou · Chunguan · Zongbo); "National construction under the hands of craftsmen...presents Altar of Land and Grain in the west, and Ancestral Temple in the east." (Rites of Zhou · Dongguan · Kao Gong Ji); "Build memorial tablets of the state, with Altar of Land and Grain in the east, and Ancestral Temple in the west." (Book of Rites · Connotations of Rituals); etc.

② JIANG Bo. Research on Ceremonial Architecture in Capitals of Han and Tang Dynasties[M]. Beijing: Cultural Relics Publishing House, 2003.

③ Chronicles of Taizu of Ming Dynasty, Volume 21.

④ Chronicles of Taizu of Ming Dynasty, Volume 24.

⑤ A Collection of Etiquettes of the Ming Dynasty, Volume 8.

⑥ Chronicles of Taizu of Ming Dynasty, Volume 25.

⑦ History of the Ming Dynasty · Ritual System, Volume 51.

⑧ Chronicles of Taizu of Ming Dynasty, Volume 100.

⑨ History of the Ming Dynasty · Ritual System, Volume 51.

⑩ Chronicles of Taizu of Ming Dynasty, Volume 60.

⑪ History of the Ming Dynasty · Ritual System, Volume 51.

⑫ Chronicles of Taizu of Ming Dynasty, Volume 114.

⑬ Total Annals of the Ming Dynasty, Volume 1.

⑭ Total Annals of the Ming Dynasty, Volume 1.

⑮ Local Chronicles of Beijing in the Ming Dynasty, Volume 17.

⑯ CAO Peng. Study on Temple and Altar Architecture in the Ming Dynasty. Tianjin: doctoral dissertation of Tianjin University, 2011.

⑰ Chronicles of Yingzong of Ming Dynasty, Volume 135.

⑱ Chronicles of Yingzong of Ming Dynasty, Volume 277.

⑲ Chronicles of Xiaozong of Ming Dynasty, Volume 1.

⑳ Chronicles of Xiaozong of Ming Dynasty, Volume 7.

㉑ Chronicles of Wuzong of Ming Dynasty, Volume 185.

㉒ Beijing Zhongshan Park Administration. Chronicle of Beijing Zhongshan Park. Beijing: China Forestry Press, 2002.

㉓ Chronicles of Wuzong of Ming Dynasty, Volume 109.

㉔ Chronicles of Shizong of Ming Dynasty, Volume 39.

㉕ Chronicles of Shizong of Ming Dynasty, Volume 51.

㉖ Chronicles of Shizong of Ming Dynasty, Volume 74.

㉗ Chronicles of Shizong of Ming Dynasty, Volume 130.

㉘ Chronicles of Shizong of Ming Dynasty, Volume 166.

㉙ Chronicles of Shizong of Ming Dynasty, Volume 167.

㉚ Chronicles of Shizong of Ming Dynasty, Volume 171.

㉛ Chronicles of Shizong of Ming Dynasty, Volumes 191, 192, 193.

㉜ Collected Statutes of the Ming Dynasty, Volume 89.

㉝ CAO Peng. Study on Temple and Altar Architecture in the Ming Dynasty. Tianjin: doctoral dissertation of Tianjin University, 2011.

㉞ Chronicles of Shizong of Ming Dynasty, Volume 216.

㉟ Chronicles of Shizong of Ming Dynasty, Volume 248.

㊱ Chronicles of Shizong of Ming Dynasty, Volume 280.

㊲ Chronicles of Shizong of Ming Dynasty, Volume 300.

㊳ CAO Peng. Study on Temple and Altar Architecture in the Ming Dynasty. Tianjin: doctoral dissertation of Tianjin University, 2011.

㊴ Chronicles of Shizu of Qing Dynasty, Volume 8.

㊵ Encyclopedia of Qing Dynasty, Volume 45.

㊶ The First Historical Archives of China. Reports for Charge-offs, 267 books in total.

㊷ YAN Kai. Research on the Architecture of Imperial Ancestral Temple in Beijing. Tianjin: master dissertation of Tianjin University, 2004.

㊸ Cases of Collected Statutes of the Ming Dynasty, Volume 427.

㊹ Beijing Zhongshan Park Administration. Chronicle of Beijing Zhongshan Park. Beijing: China Forestry Press, 2002.

㊺ Beijing Zhongshan Park Administration. Chronicle of Beijing Zhongshan Park. Beijing: China Forestry Press, 2002.

㊻ Architectural Design and Research Institute of Tianjin University, Beijing Working People's Cultural Palace. Protection Planning of Ancestral Temple as a key cultural relics unit under national protection, 2014.

㊼ Beijing Zhongshan Park Administration. Chronicle of Beijing Zhongshan Park. Beijing: China Forestry Press, 2002.

㊽ Architectural Design and Research Institute of Tianjin University, Beijing Working People's Cultural Palace. Protective planning of Ancestral Temple as a key cultural relics unit under national protection, 2014.

㊾ Beijing Zhongshan Park Administration. Chronicle of Beijing Zhongshan Park. Beijing: China Forestry Press, 2002.

㊿ ZHANG Guorui. Ancestral Temple Study. Beijing: Palace Museum, 1932.

51 See: JIANG Shunyuan. Ancestral Temple System in the Qing Dynasty. Journal of the Palace Museum, 1987(3). 15-23+57; Fu Gongyue. An-

cestral Temple of the Qing Dynasty. Journal of the Palace Museum, 1986(3). 73-79+86.

㉜ See: YAN Kai. Research on the Architecture of Imperial Ancestral Temple in Beijing. Tianjin: master dissertation of Tianjin University, 2004; YA Baiyang. Research on Architecture of Beijing Altar of Land and Grain. Tianjin: master dissertation of Tianjin University, 2005; CHI Xiaoyan, CAO Peng. Discrimination of Targets of Worship of Beijing Temple of Earth and Altar of Land and Grain - Outline on Evolution of Gods of Earth as well as Land and Grain Through Historical Periods. Journal of Shenyang Jianzhu University (Social Science Edition), 2006(4), 309-312; YAN Kai, WANG Qiheng, CAO Peng. Comparative Study on the Three Royal Palaces of the Ming and Qing Dynasties in Beijing. Journal of Shandong Institute of Architecture & Engineering, 2006(2), 116-121; YA Baiyang. Intent of Space Layout Design of the Altar of Land and Grain in Ming and Qing Dynasties - Exploring the Origin of the "temple in east and altar in west" Layout. Traditional Chinese Architecture and Gardens, 2010(2), 57-59; CAO Peng. Study on Temple and Altar Architecture in the Ming Dynasty. Tianjin: doctoral dissertation of Tianjin University, 2011; etc.

㉝ Architectural Design and Research Institute of Tianjin University, Beijing Working People's Cultural Palace. Protection Planning of Ancestral Temple as a key cultural relics unit under national protection, 2014.

㉞ Emperor Authorized Collected Statutes of the Great Qing Dynasty, Volume 40.

㉟ GUO Huayu. Study on Carpentry Work of Official Architecture in the Ming Dynasty[M]. Nanjing: Southeast University Press. 2005.

㊱ Emperor Authorized Collected Statutes of the Great Qing Dynasty, Volume 71.

㊲ Emperor Authorized Collected Statutes of the Great Qing Dynasty, Volume 43.

㊳ Total Annals of the Ming Dynasty, Volume 1.

㊴ Emperor Authorized Collected Statutes of the Great Qing Dynasty, Volume 71.

㊵ Collected Statutes of the Great Ming Dynasty, Volume 85.

㊶ Chronicles of Gaozong (posthumous title: chun, meaning purity), Volume 921.

㊷ Emperor Authorized Collected Statutes of the Great Qing Dynasty, Volume 71.

㊸ CAO Peng. Study on Temple and Altar Architecture in the Ming Dynasty[D]. Tianjin: doctoral dissertation of Tianjin University, 2011.

㊹ Chronicles of Taizu of Ming Dynasty, Volume 114.

㊺ Beijing Zhongshan Park Administration. Chronicle of Beijing Zhongshan Park. Beijing: China Forestry Press, 2002.

㊻ QI Yingtao. On Shape and Structure of Beijing Palace-like Wooden Construction of the Ming Dynasty[M]//. QI Yingtao. Ancient Architecture Papers by QI Yingtao. Beijing: Huaxia Publishing House, 1992: 325-341.

图
版

Drawings

太庙

The Ancestral Temple

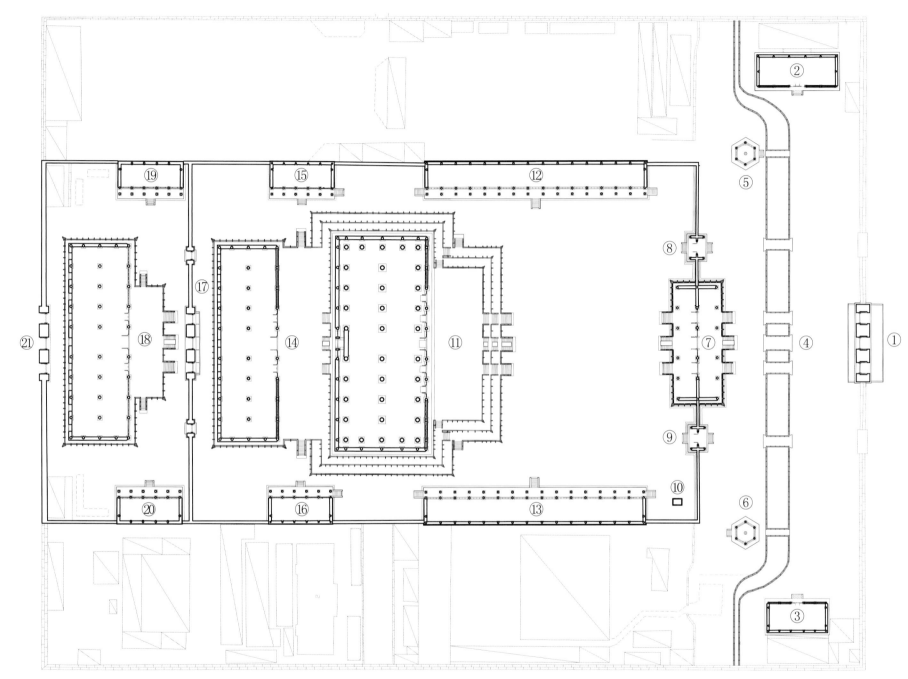

1　头道门　The First Gate
2　神库　The Divine Store
3　神厨　The Divine Kitchen
4　金水桥　The Golden Water Bridge
5　东井亭　The East Well Pavilion
6　西井亭　The West Well Pavilion
7　大戟门　The Daji Gate
8　东小戟门　The East Xiaoji Gate
9　西小戟门　The West Xiaoji Gate
10　西焚帛炉　The West Silk Burner
11　前殿　The Front Hall
12　前殿东配殿　The East Side Hall in Front
13　前殿西配殿　The West Side Hall in Front
14　中殿　The Central Hall
15　中殿东配殿　The East Side Hall of the Central Hall
16　中殿西配殿　The West Side Hall of the Central Hall
17　二道门　The Second Gate
18　后殿　The Rear Hall
19　后殿东配殿　The East Side Hall of the Rear Hall
20　后殿西配殿　The West Side Hall of the Rear Hall
21　三道门　The Third Gate

太庙第二进院落总平面图
General Plan of the Second Courtyard of the Ancestral Temple

0　　10　　20m

# 头道门
## The First Gate

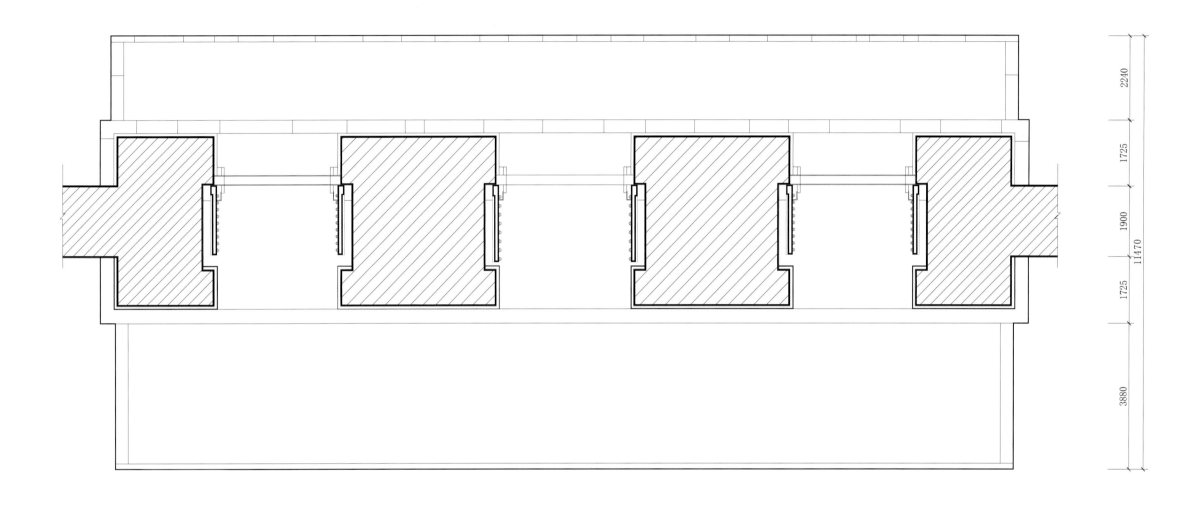

太庙头道门平面图
Plan of the First Gate of the Ancestral Temple

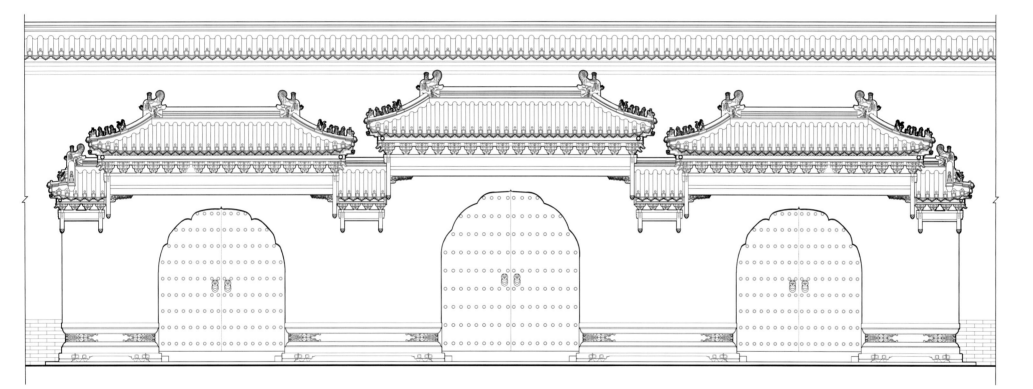

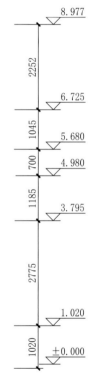

太庙头道门南立面图
South elevation of the First Gate to the Ancestral Temple

0  1  3m

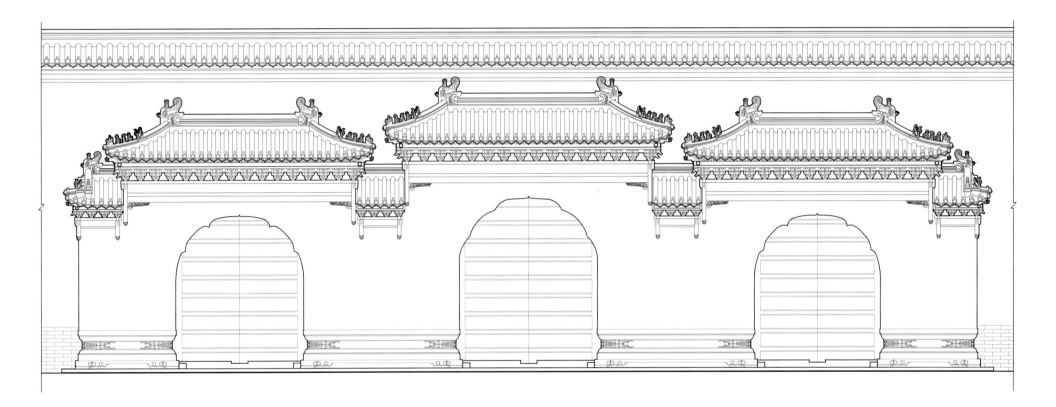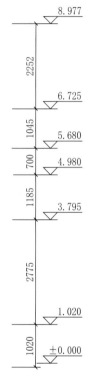

8.977

2252

6.725

1045

5.680

700

4.980

1185

3.795

2775

1.020

1020

±0.000

太庙头道门北立面图
North elevation of the First Gate to the Ancestral Temple

0　　1　　　3m

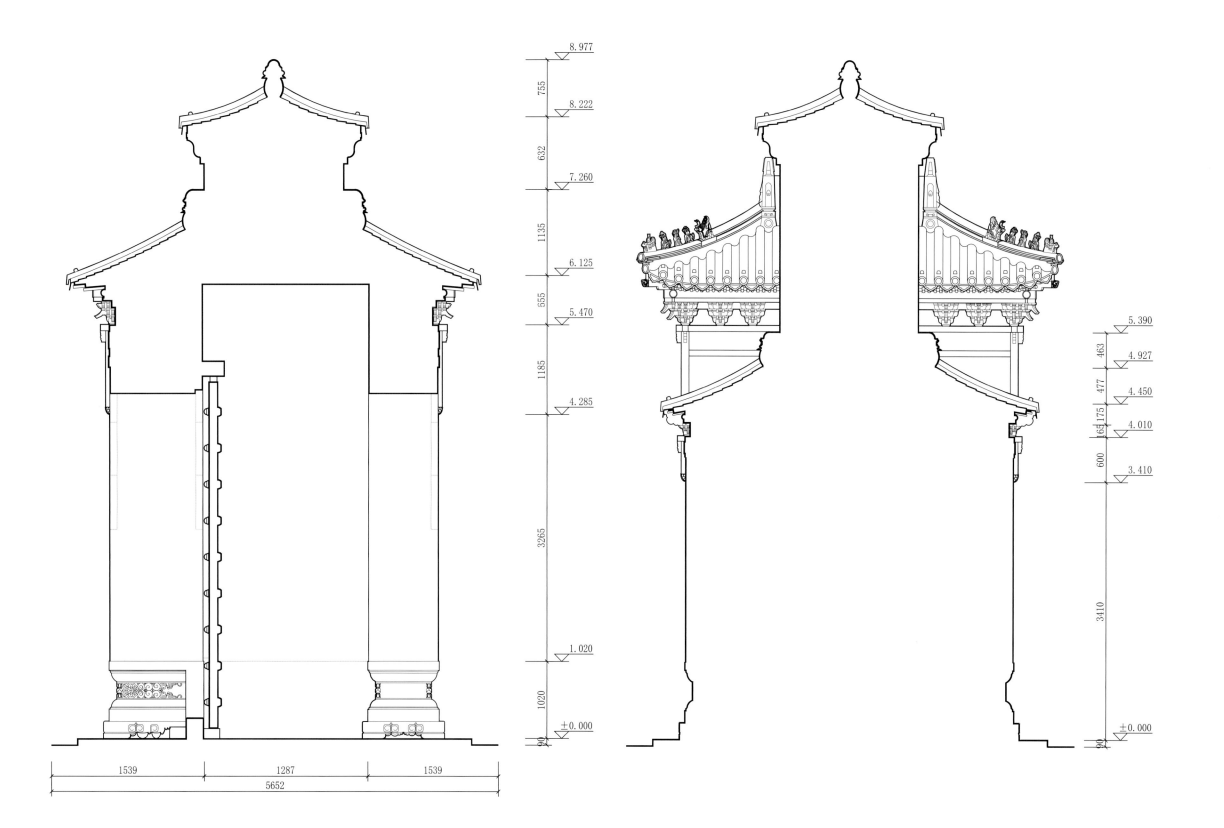

8.977

755

8.222

632

7.260

1135

6.125

555

5.470

1185

4.285

3265

1.020

1020

±0.000

90

1539　　1287　　1539

5652

5.390

463

4.927

477

4.450

175

4.010

63

600

3.410

3410

±0.000

90

太庙头道门中门剖面图

Cross-section of the middle entrance of the First Gate to the Ancestral Temple

太庙头道门中门侧立面图

Side elevation of the middle entrance of the First Gate to the Ancestral Temple

0　　　　1　　　　2m

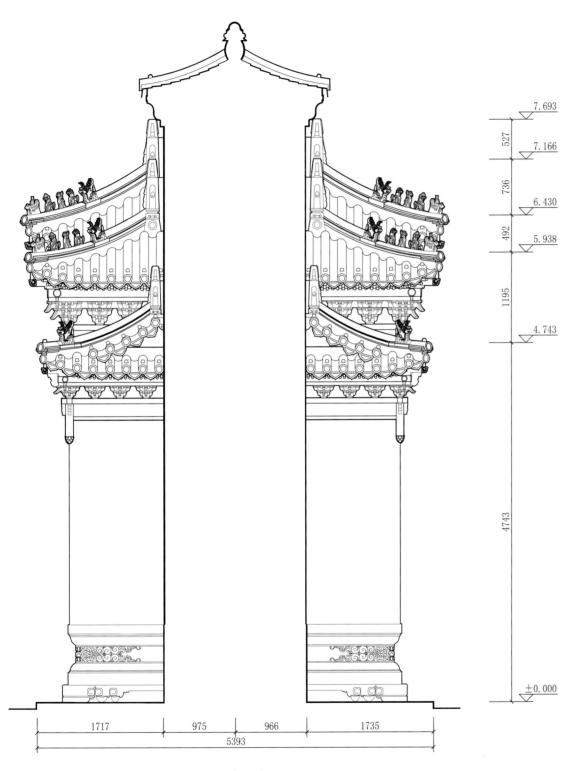

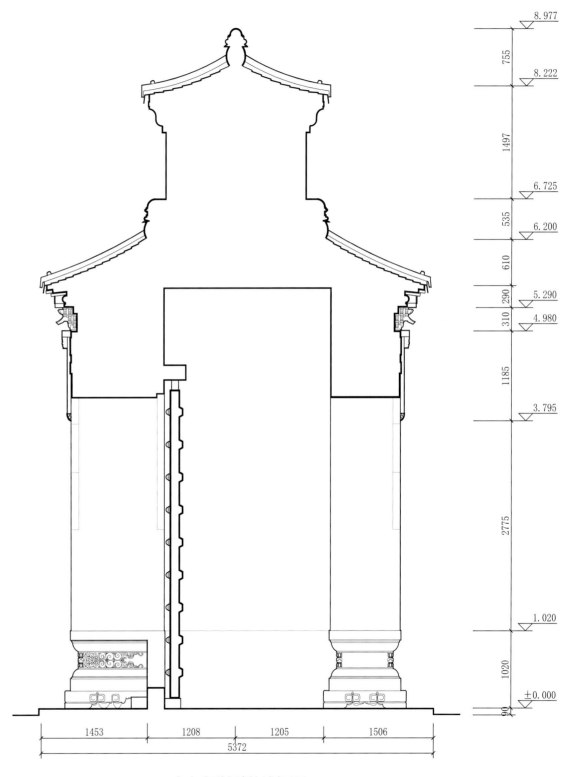

太庙头道门侧门侧立面图
Side elevation of the side entrance of the First Gate to the Ancestral Temple

太庙头道门侧门剖面图
Cross-section of the side entrance of the First Gate to the Ancestral Temple

0    1    2m

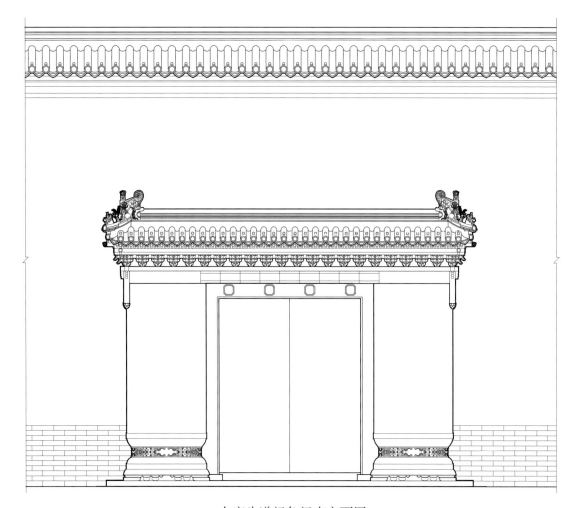

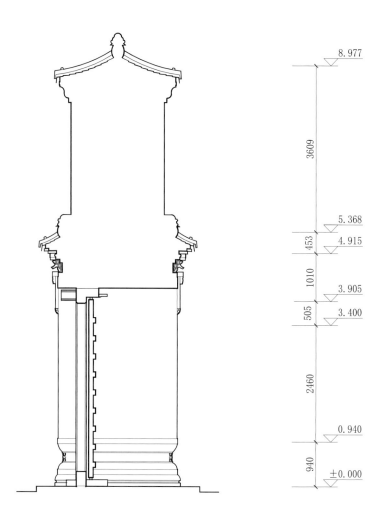

8.977

3609

5.368

453 4.915

1010 3.905

505 3.400

2460

0.940

940 ±0.000

太庙头道门角门南立面图

South elevation of the corner entrance of the First Gate to the Ancestral Temple

太庙头道门角门剖面图

Cross-section of the corner entrance of the First Gate to the Ancestral Temple

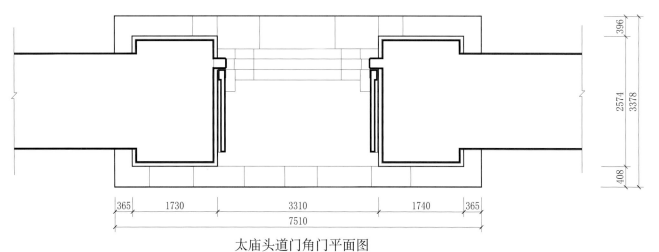

396

2574 3378

408

365 1730 3310 1740 365

7510

太庙头道门角门平面图

Plan of the corner entrance of the First Gate to the Ancestral Temple

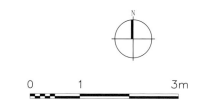

0    1         3m

# 金水桥
# The Golden
# Water Bridge

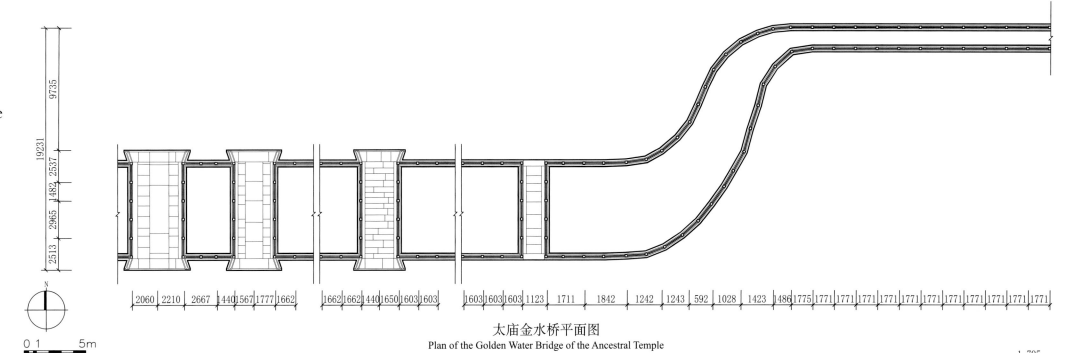

太庙金水桥平面图
Plan of the Golden Water Bridge of the Ancestral Temple

0 1 5m

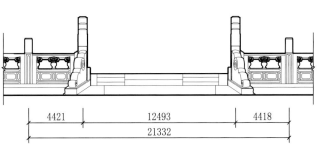

太庙金水桥一号桥正立面
Front elevation of Bridge One of the Golden Water Bridge of the Ancestral Temple

4421　12493　4418
21332

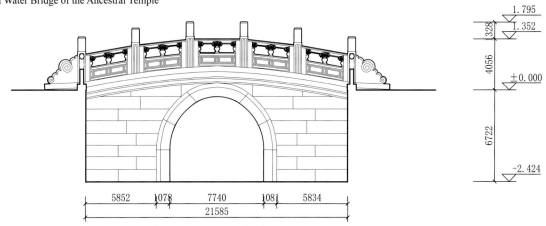

太庙金水桥一号桥侧立面图
Side elevation of Bridge One of the Golden Water Bridge of the Ancestral Temple

5852　1078　7740　1081　5834
21585

1.795
1.352
±0.000
-2.424

0 1 3m

太庙金水桥一号桥横剖面图
Cross-section of Bridge One of the Golden Water Bridge of the Ancestral Temple

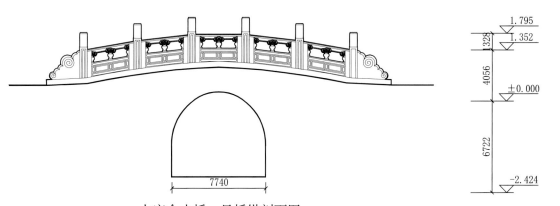

太庙金水桥一号桥纵剖面图
Longitudinal Section of Bridge One of the Golden Water Bridge of the Ancestral Temple

7740

1.795
1.352
±0.000
-2.424

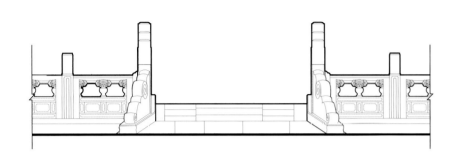

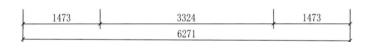

太庙金水桥二号桥正立面图

Front elevation of Bridge Two of the Golden Water Bridge of the Ancestral Temple

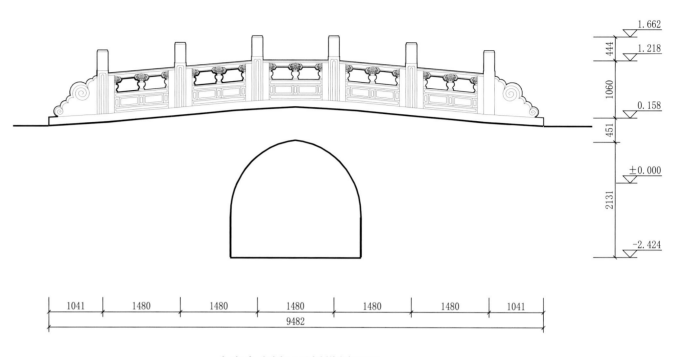

太庙金水桥二号桥纵剖面图

Longitudinal sections of Bridge Two of the Golden Water Bridge of the Ancestral Temple

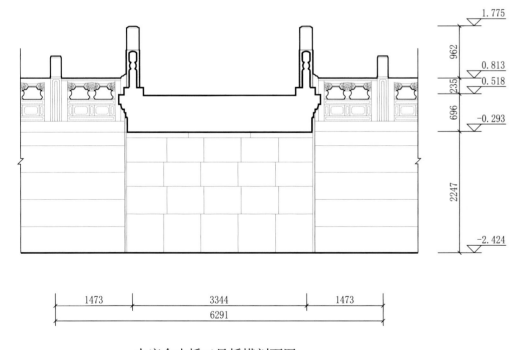

太庙金水桥二号桥横剖面图

Cross-sections of Bridge Two of the Golden Water Bridge of the Ancestral Temple

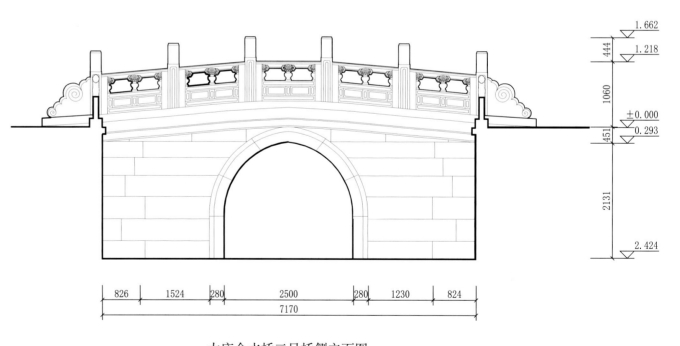

太庙金水桥二号桥侧立面图

Side elevation of Bridge Two of the Golden Water Bridge of the Ancestral Temple

0　　　1　　　　　　　3m

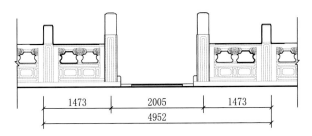

太庙金水桥四号桥正立面图
Front elevation of Bridge Four of the Golden Water Bridge of the Ancestral Temple

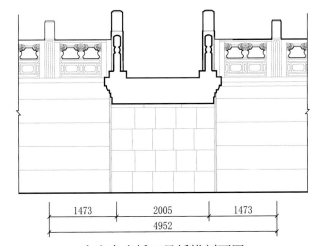

太庙金水桥四号桥横剖面图
Cross sections of Bridge Four of the Golden Water Bridge of the Ancestral Temple

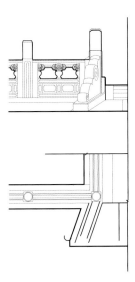

太庙金水桥细部
Details of the Golden Water Bridge of the Ancestral Temple

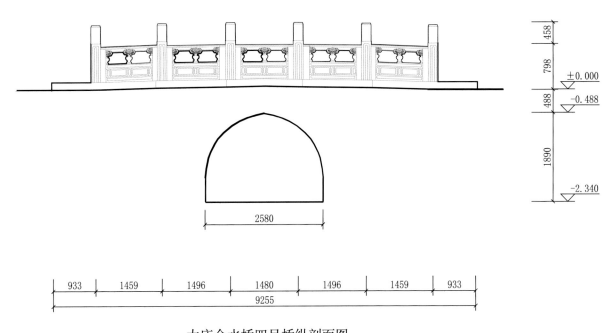

太庙金水桥四号桥纵剖面图
Longitudinal sections of Bridge Four of the Golden Water Bridge of the Ancestral Temple

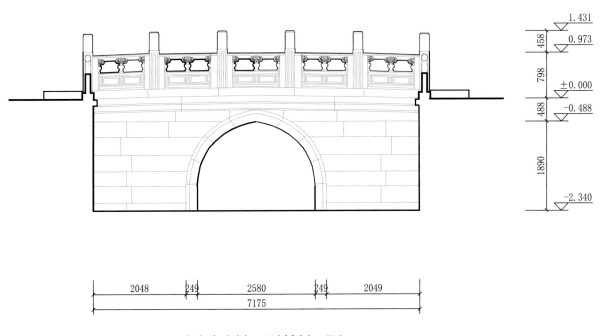

太庙金水桥四号桥侧立面图
Side elevation of Bridge Four of the Golden Water Bridge of the Ancestral Temple

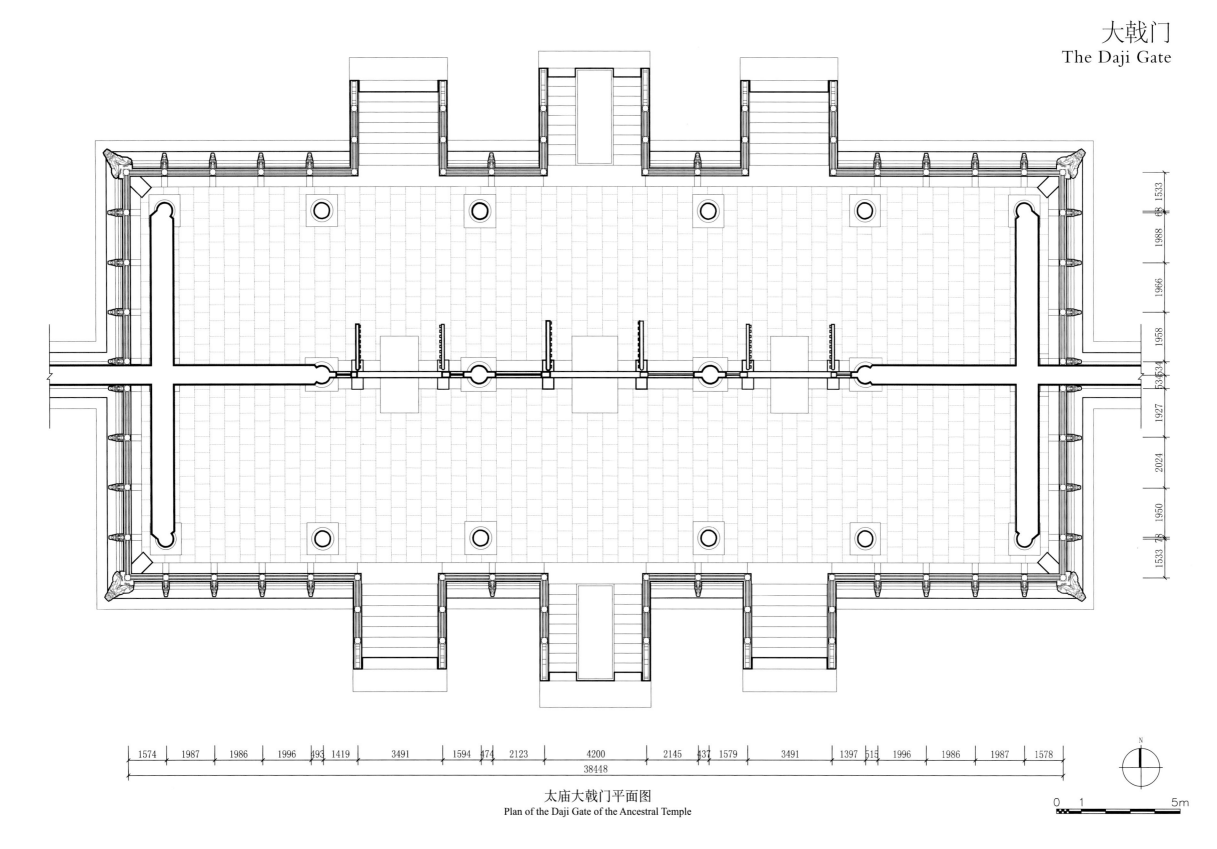

中国古建筑测绘大系·坛庙建筑 —— 太庙和社稷坛

048

| 1533 | 1988 | 1966 | 1958 | 1534 | 1927 | 2024 | 1950 | 1533 |

| 1574 | 1987 | 1986 | 1996 | 493 | 1419 | 3491 | 1594 | 474 | 2123 | 4200 | 2145 | 437 | 1579 | 3491 | 1397 | 515 | 1996 | 1986 | 1987 | 1578 |

38448

太庙大戟门平面图
Plan of the Daji Gate of the Ancestral Temple

N

0 1 5m

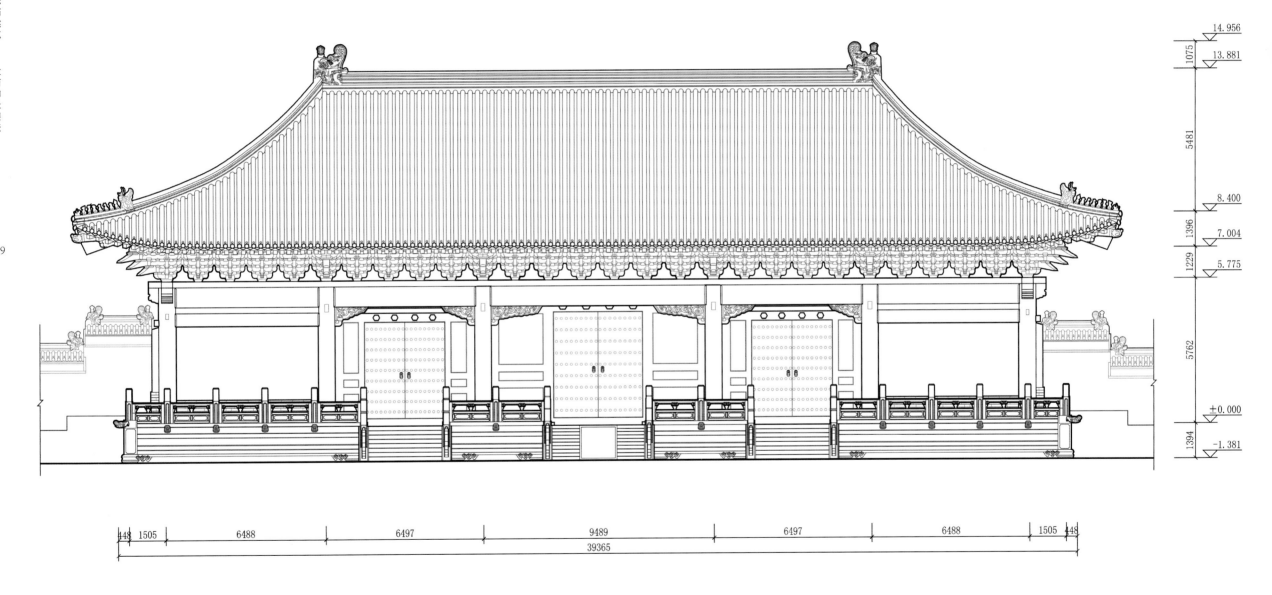

14.956
13.881
1075

5481

8.400
7.004
1396

5.775
1229

5762

±0.000

1394
-1.381

448 1505    6488    6497    9489    6497    6488    1505 448

39365

太庙大戟门南立面图
South elevation of the Daji Gate of the Ancestral Temple

0   1    3m

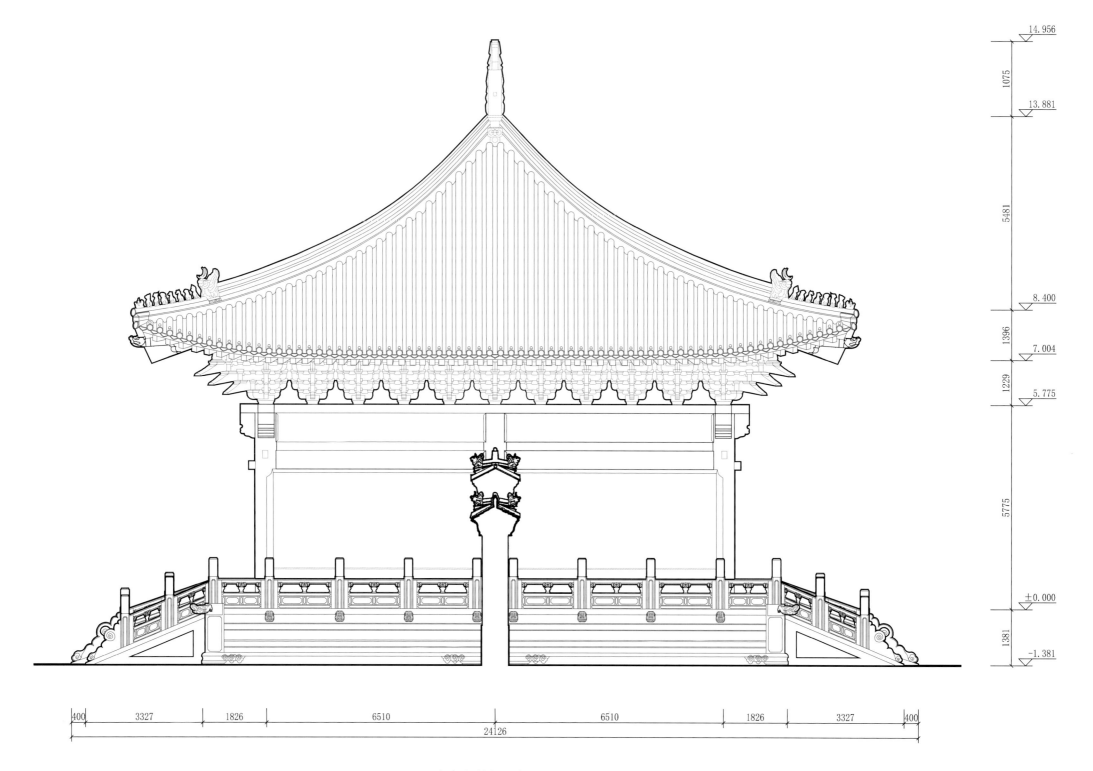

14.956

1075

13.881

5481

8.400

1396

7.004

1229

5.775

5775

±0.000

1381

-1.381

400 3327 1826 6510 6510 1826 3327 400

24126

太庙大戟门西立面图
West elevation of the Daji Gate of the Ancestral Temple

0 1 3m

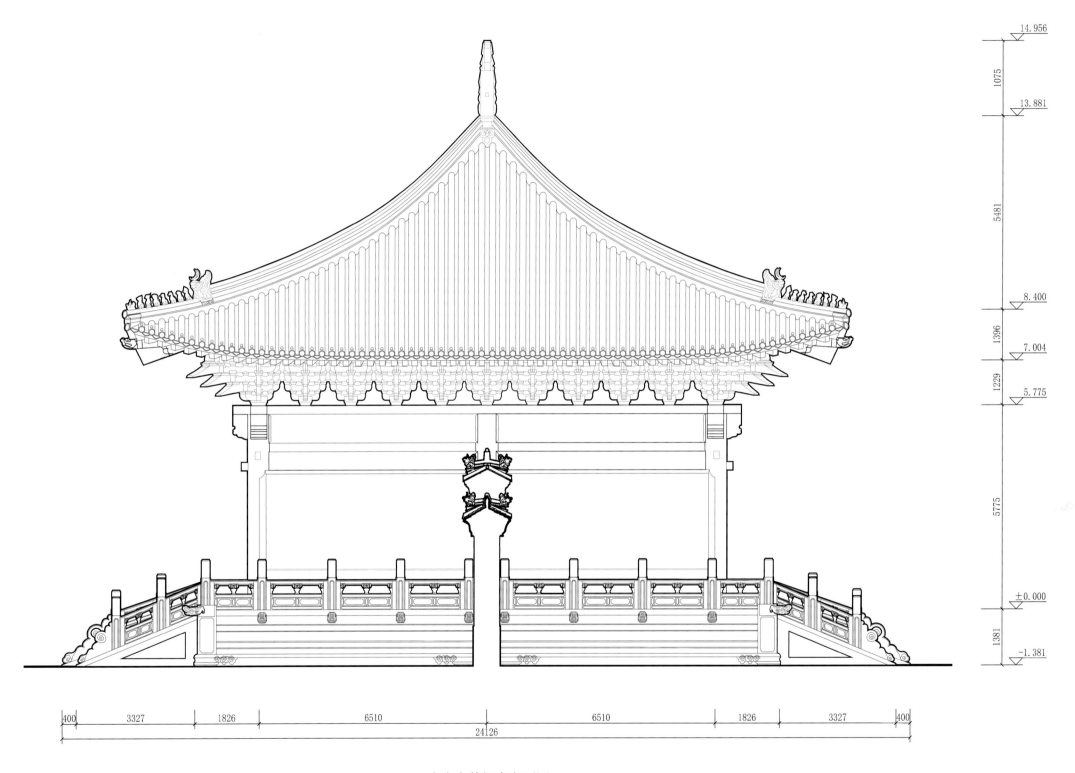

051

14.956

1075

13.881

5481

8.400

1396

7.004

1229

5.775

5775

±0.000

1381

−1.381

400  3327  1826  6510  6510  1826  3327  400

24126

太庙大戟门东立面图
East elevation of the Daji Gate of the Ancestral Temple

0  1  3m

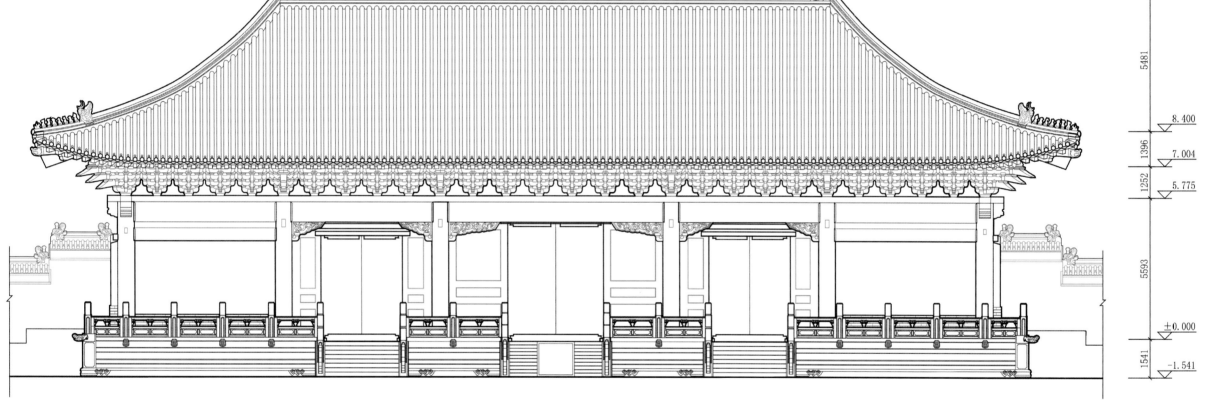

052

14.956
1075
13.881

5481

8.400
1396
7.004
1252
5.775

5593

±0.000
1541
-1.541

448 1513　6470　6442　9486　6442　6470　1513 448
39233

太庙大戟门北立面图
North elevation of the Daji Gate of the Ancestral Temple

0　1　　　5m

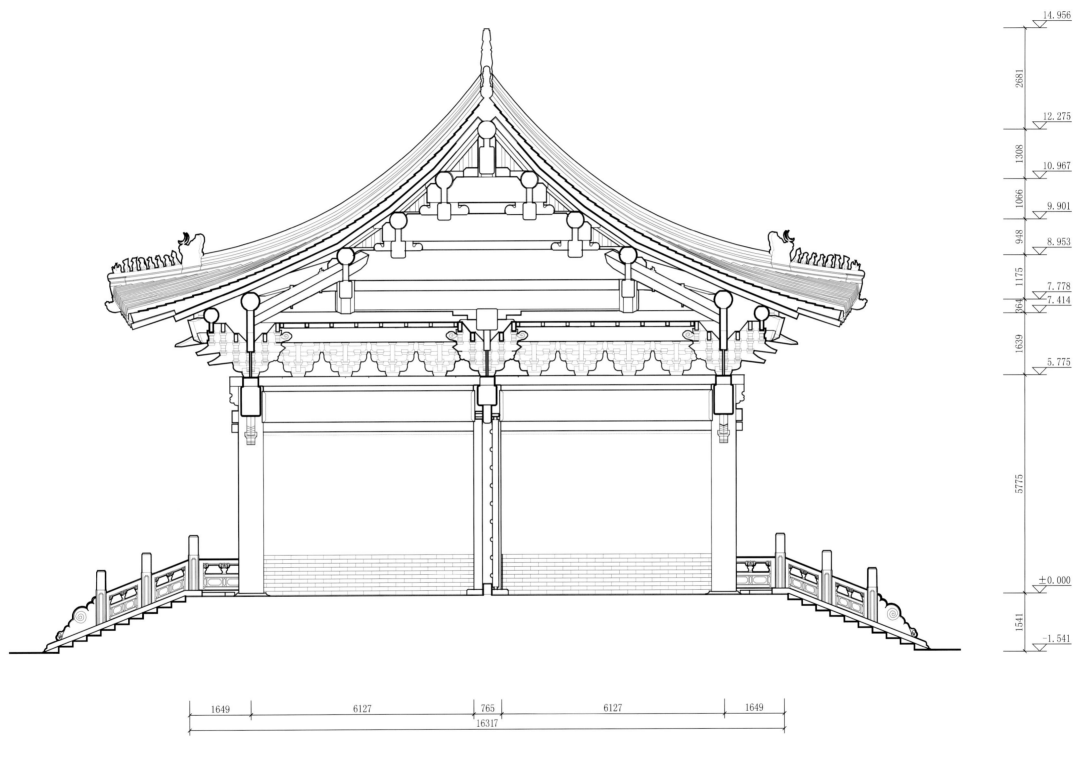

14.956

2681

12.275

1308

10.967

1066

9.901

948

8.953

1175

7.778

364

7.414

1639

5.775

5775

±0.000

1541

-1.541

1649　　6127　　765　　6127　　1649

16317

太庙大戟门明间剖面图
Section of the central bay of the Daji Gate of the Ancestral Temple

0　　1　　3m

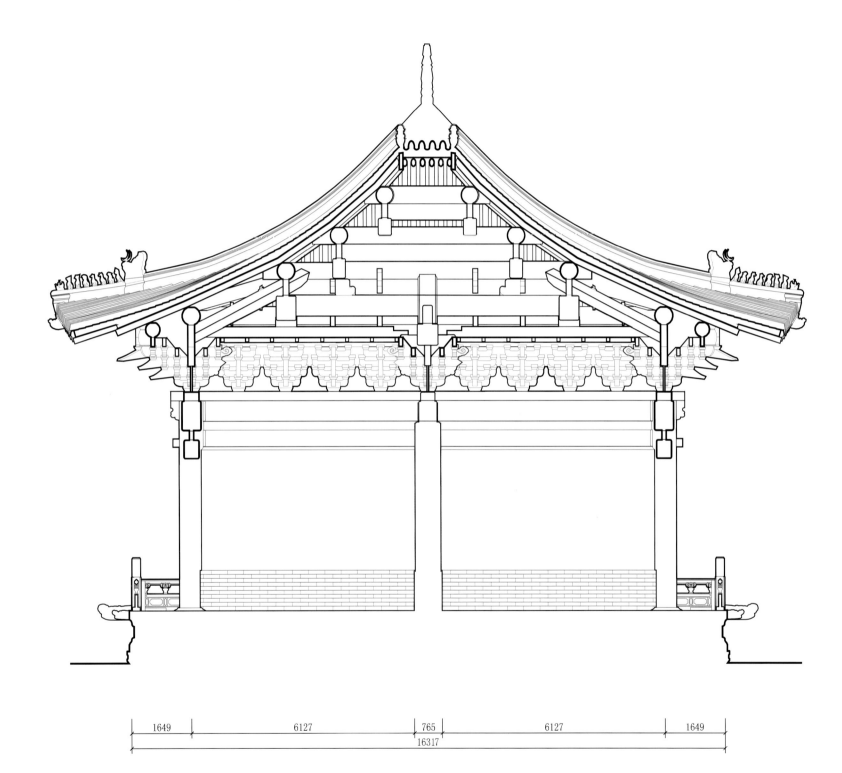

14. 956

2681

12. 275

1308

10. 967

1066

9. 901

948

8. 953

1175

7. 778

364
7. 414

1639

5. 775

5775

±0. 000

1541

−1. 541

1649　　6127　　765　　6127　　1649

16317

太庙大戟门梢间剖面图

Section of the terminal bay of the Daji Gate of the Ancestral Temple

0　　1　　3m

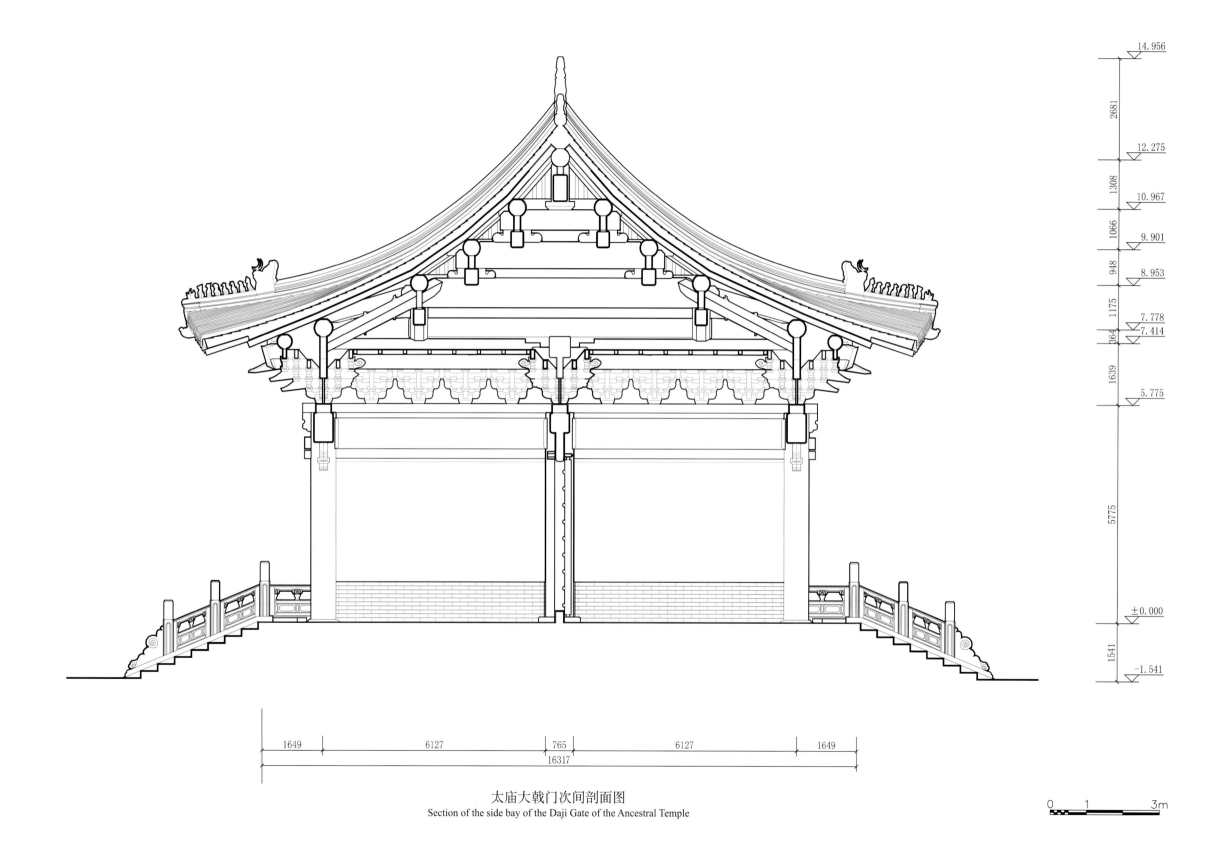

14.956

2681

12.275

1308

10.967

1066

9.901

948

8.953

1175

7.778
7.414

364

1639

5.775

5775

±0.000

1541

-1.541

1649　6127　765　6127　1649

16317

太庙大戟门次间剖面图

Section of the side bay of the Daji Gate of the Ancestral Temple

0　1　　　3m

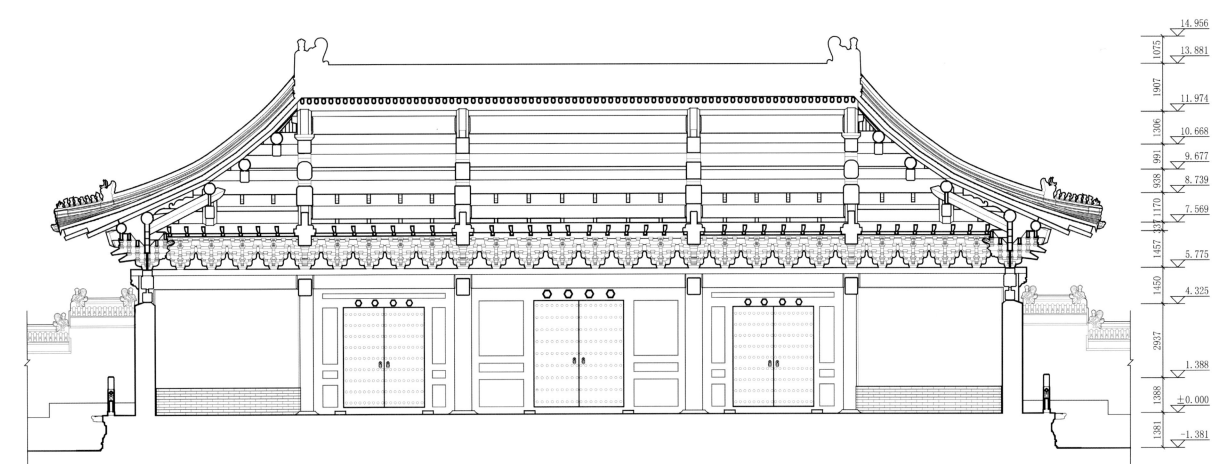

14.956
1075
13.881
1907
11.974
1306
10.668
991
9.677
938
8.739
337 1170
7.569
1457
5.775
1450
4.325
2937
1.388
1388
±0.000
1381
-1.381

448 1505　6488　6497　9489　6497　6488　1505 448
39365

太庙大戟门纵剖面图
Longitudinal section of the Daji Gate of the Ancestral Temple

0　1　5m

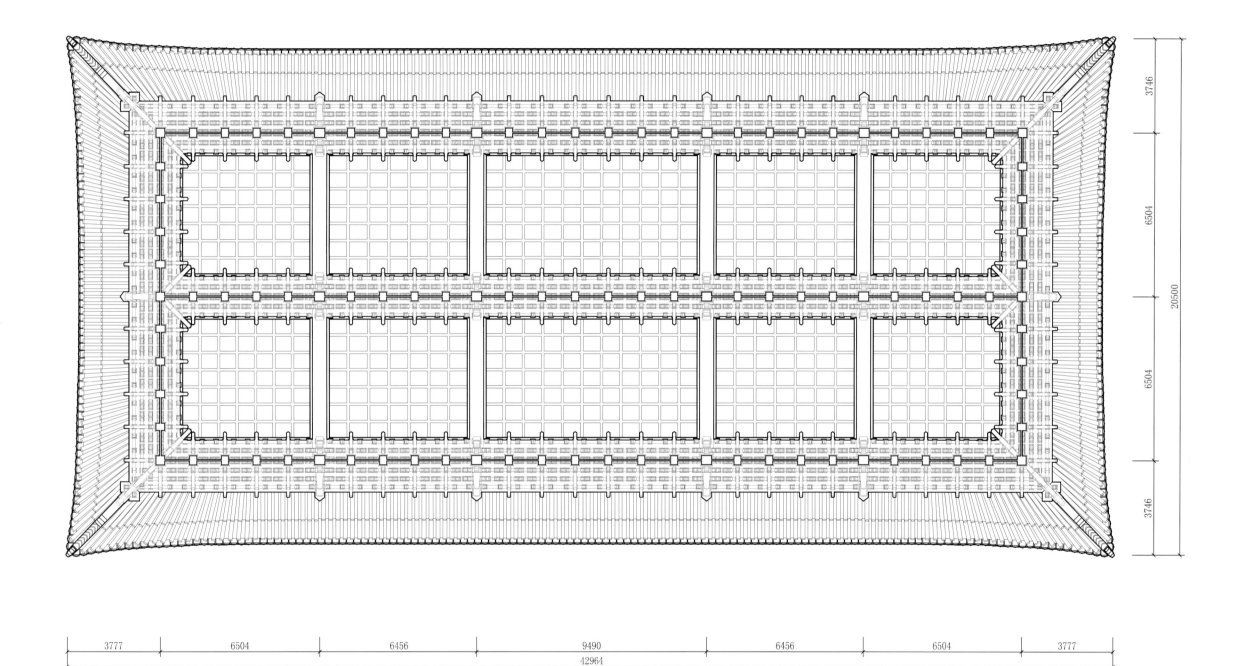

3746

6504

20500

6504

3746

3777　　6504　　6456　　9490　　6456　　6504　　3777

42964

太庙大戟门天花仰视图

Bottom View of the ceiling of the Daji Gate of the Ancestral Temple

0　1　　　　5m

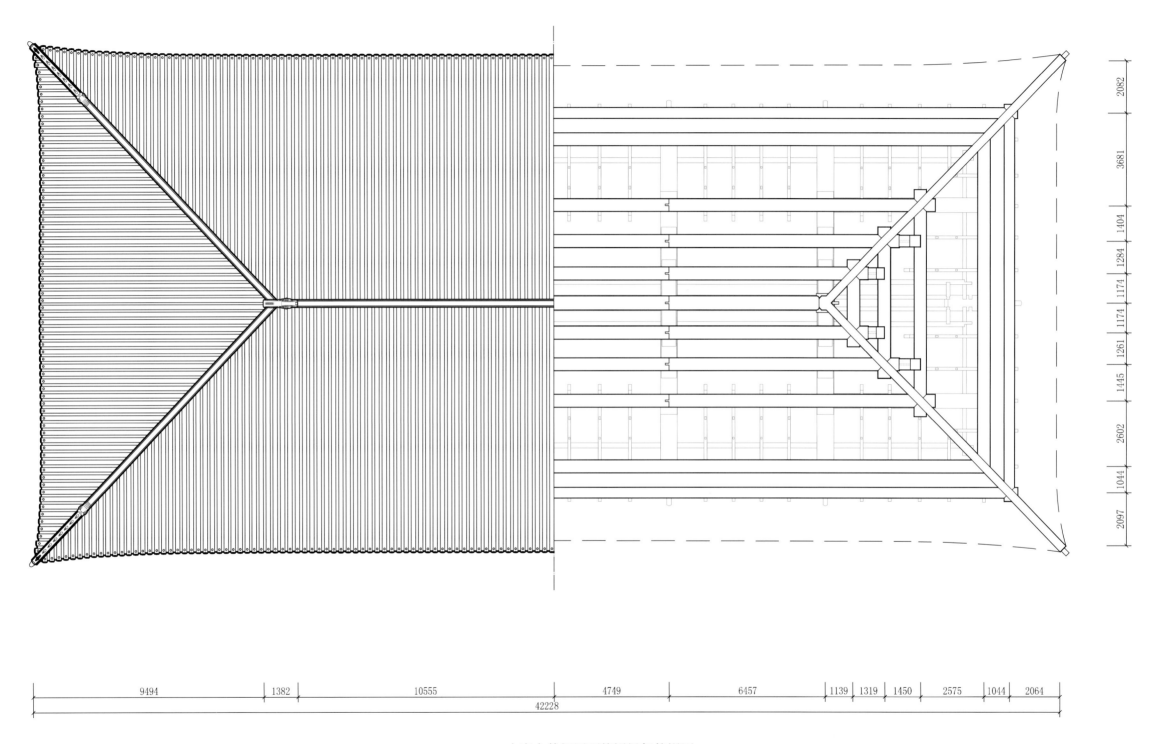

太庙大戟门屋面俯视梁架仰视图
Roof and beam frame of the Daji Gate of the Ancestral Temple

0 1 5m

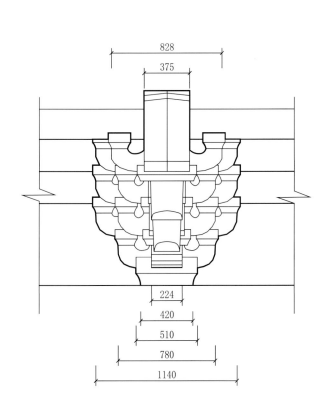

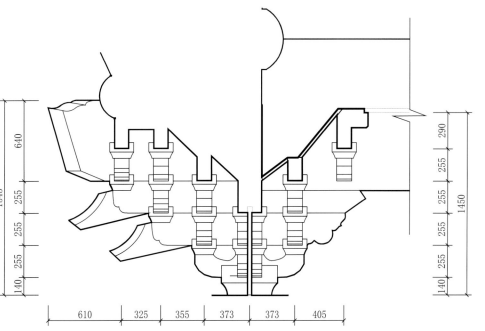

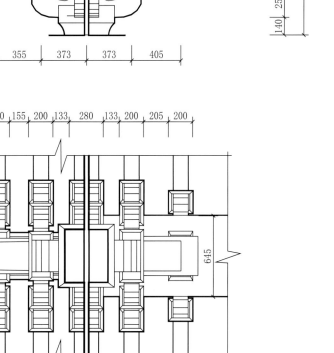

太庙大戟门外檐柱头科斗栱大样图

Detail of the connecting bracket sets on the outer columns of the Daji Gate of the Ancestral Temple

0    0.2    0.4m

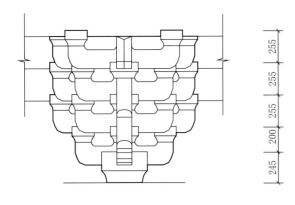

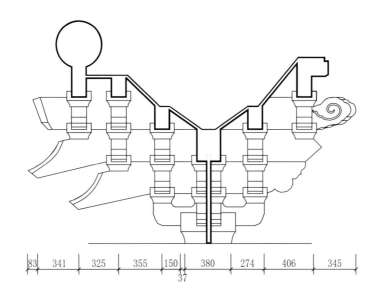

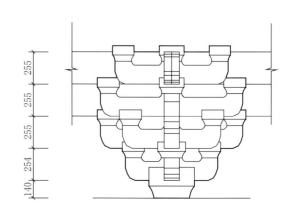

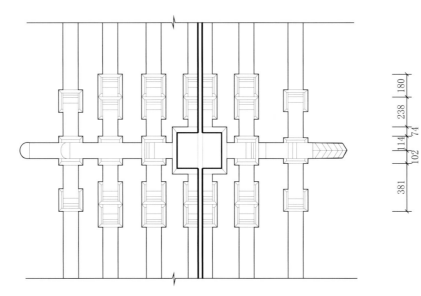

太庙大戟门外檐平身科斗栱大样图

Detail of the connecting bracket sets between the outer columns of the Daji Gate of the Ancestral Temple

0    0.3    0.6m

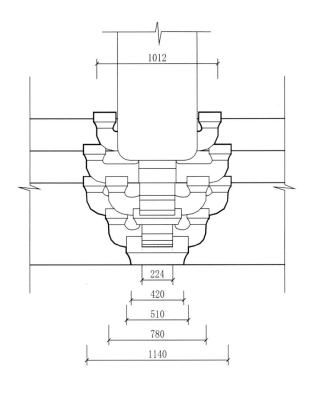

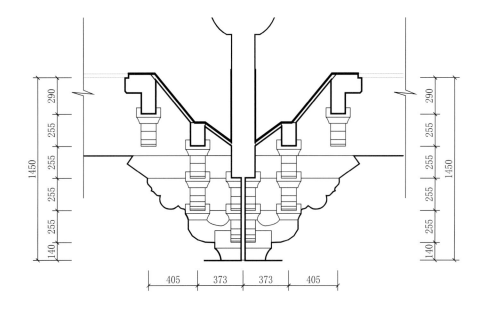

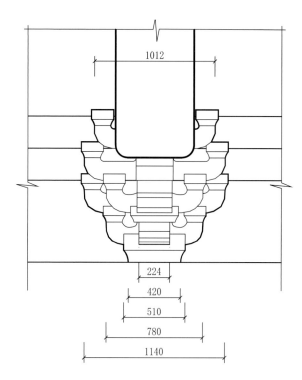

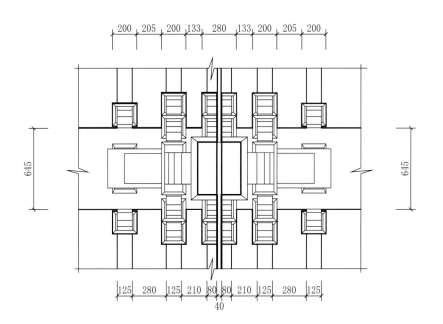

太庙大戟门内檐柱头科斗栱大样图

Detail of the connecting bracket sets on the inner columns of the Daji Gate of the Ancestral Temple

0    0.2    0.4m

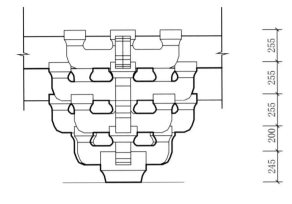

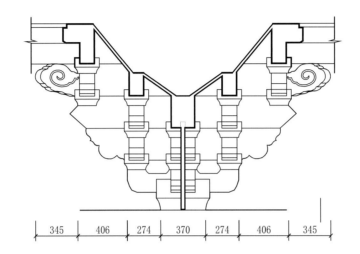

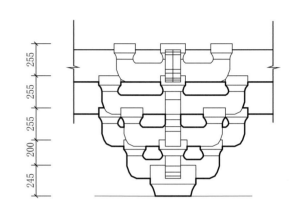

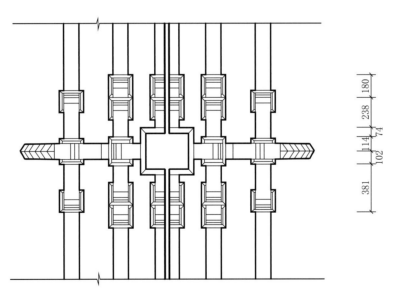

太庙大戟门内檐平身科斗栱大样图

Detail of the connecting bracket sets between the inner columns of the Daji Gate of the Ancestral Temple

0    0.2    0.4m

# 小戟门
## The Xiaoji Gate

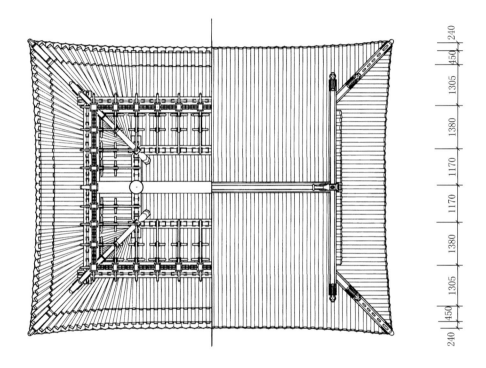

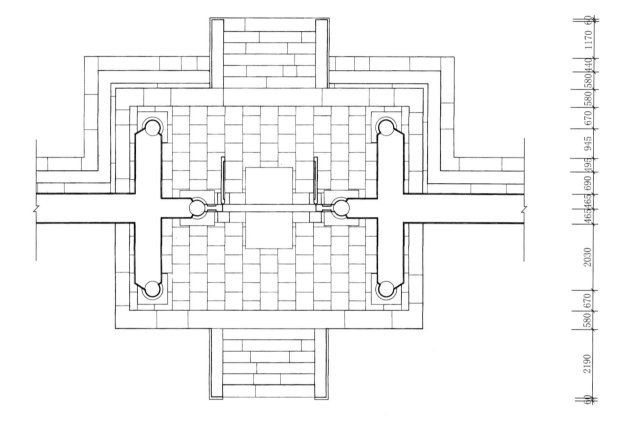

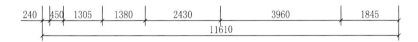

太庙小戟门梁架俯仰图
Top view of the beam frame of the Xiaoji Gate to the Ancestral Temple

太庙小戟门平面图
Plan of the Xiaoji Gate to the Ancestral Temple

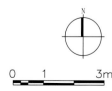

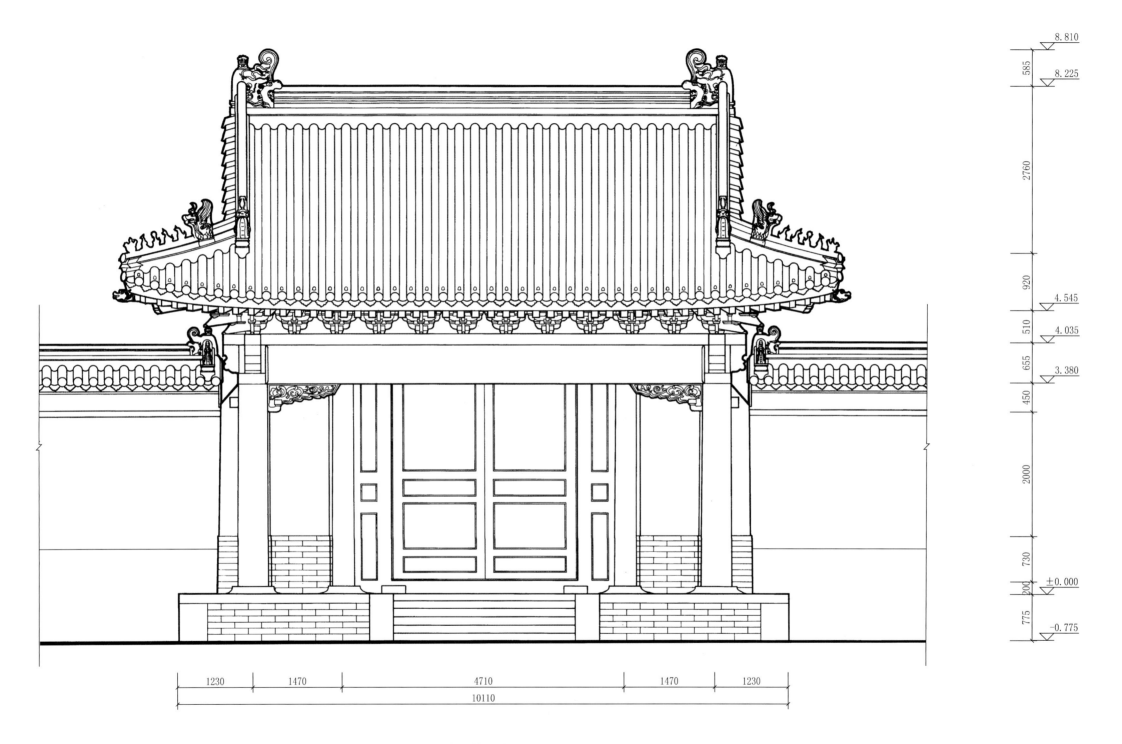

8.810
585
8.225
2760
920
4.545
510
4.035
655
3.380
450
2000
730
±0.000
200
775
-0.775

1230　1470　4710　1470　1230
10110

太庙小戟门南立面图
South elevation of the Xiaoji Gate to the Ancestral Temple

0　0.4　2m

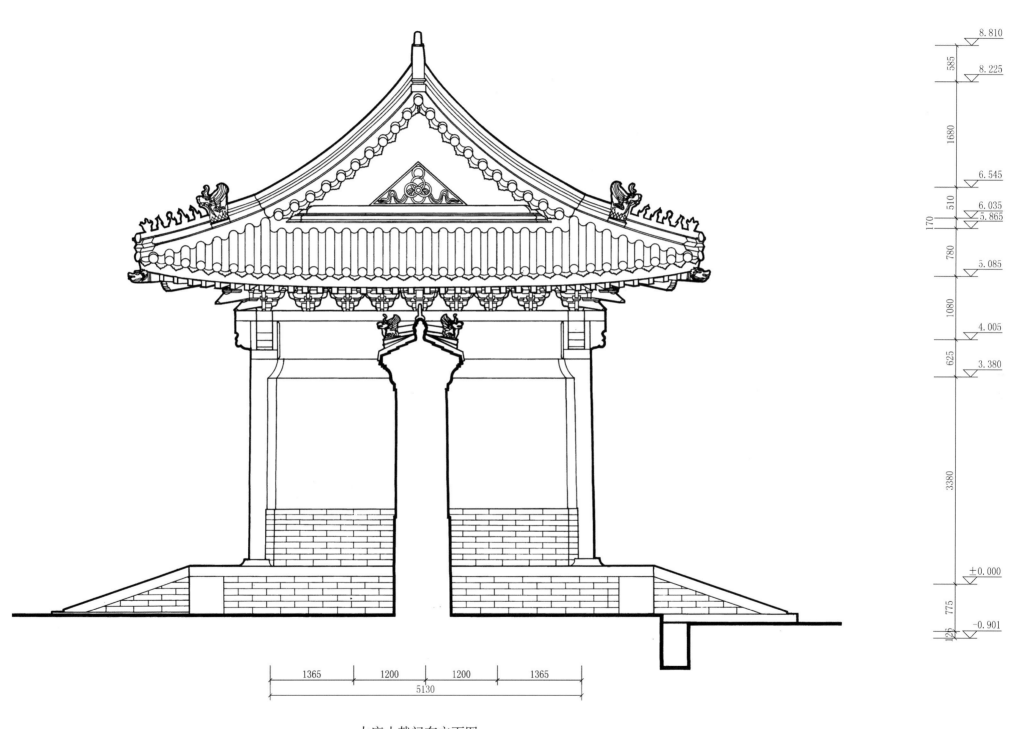

8.810

585

8.225

1680

6.545

510

6.035

170

5.865

780

5.085

1080

4.005

625

3.380

3380

±0.000

775

-0.901

125

1365  1200  1200  1365

5130

太庙小戟门东立面图
East elevation of the Xiaoji Gate to the Ancestral Temple

0    0.4         2m

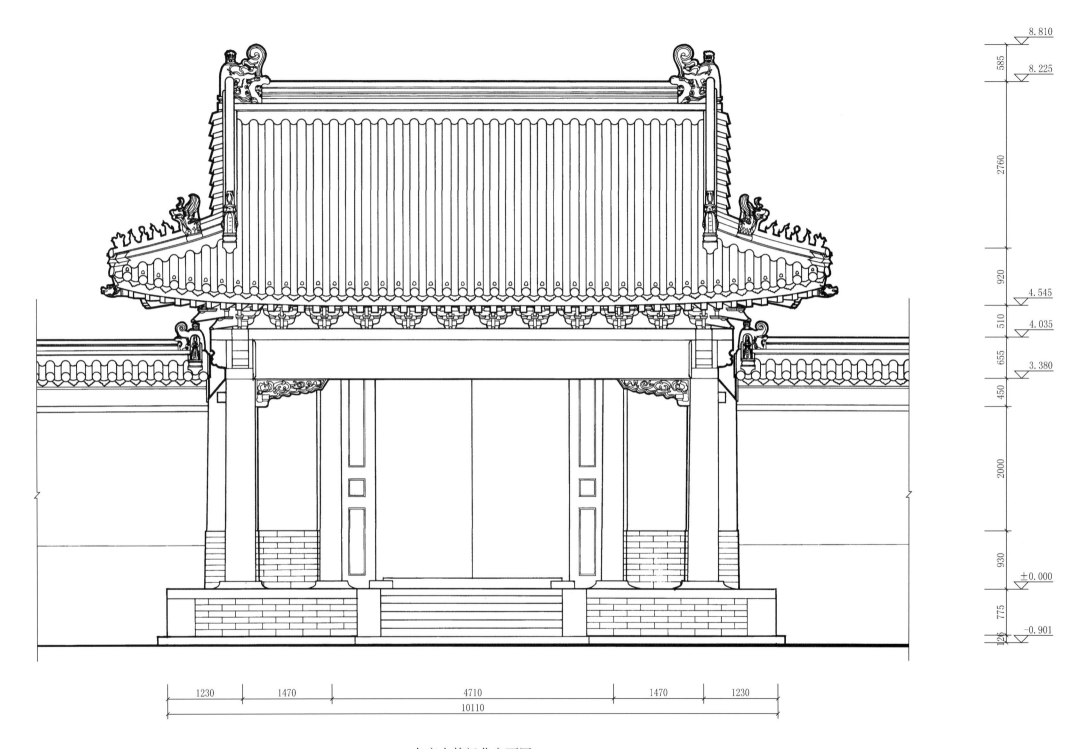

8.810

585

8.225

2760

920

4.545

510

4.035

655

3.380

450

2000

930

±0.000

775

126

-0.901

1230 1470 4710 1470 1230

10110

太庙小戟门北立面图

North elevation of the Xiaoji Gate to the Ancestral Temple

0   0.4        2m

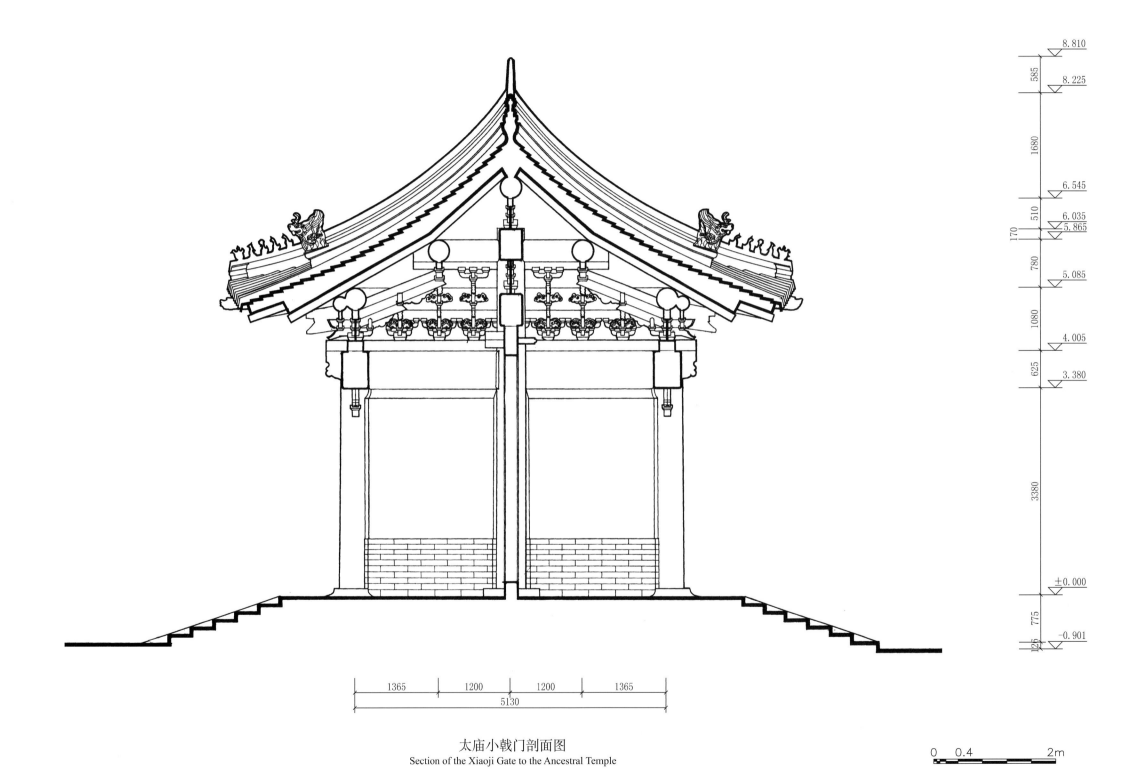

8.810
585
8.225
1680
6.545
510
6.035
170
5.865
780
5.085
1080
4.005
625
3.380
3380
±0.000
775
-0.901
125

1365　1200　1200　1365
5130

太庙小戟门剖面图
Section of the Xiaoji Gate to the Ancestral Temple

0　0.4　2m

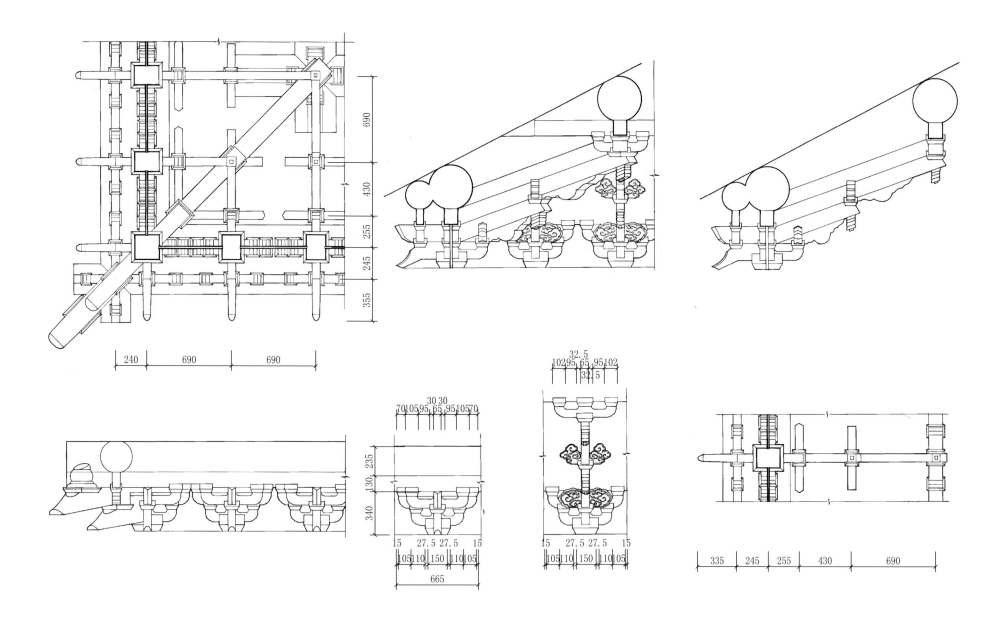

太庙小戟门斗栱大样图
Detail of the connecting bracket sets of the Xiaoji Gate of the Ancestral Temple

0    0.2    0.4m

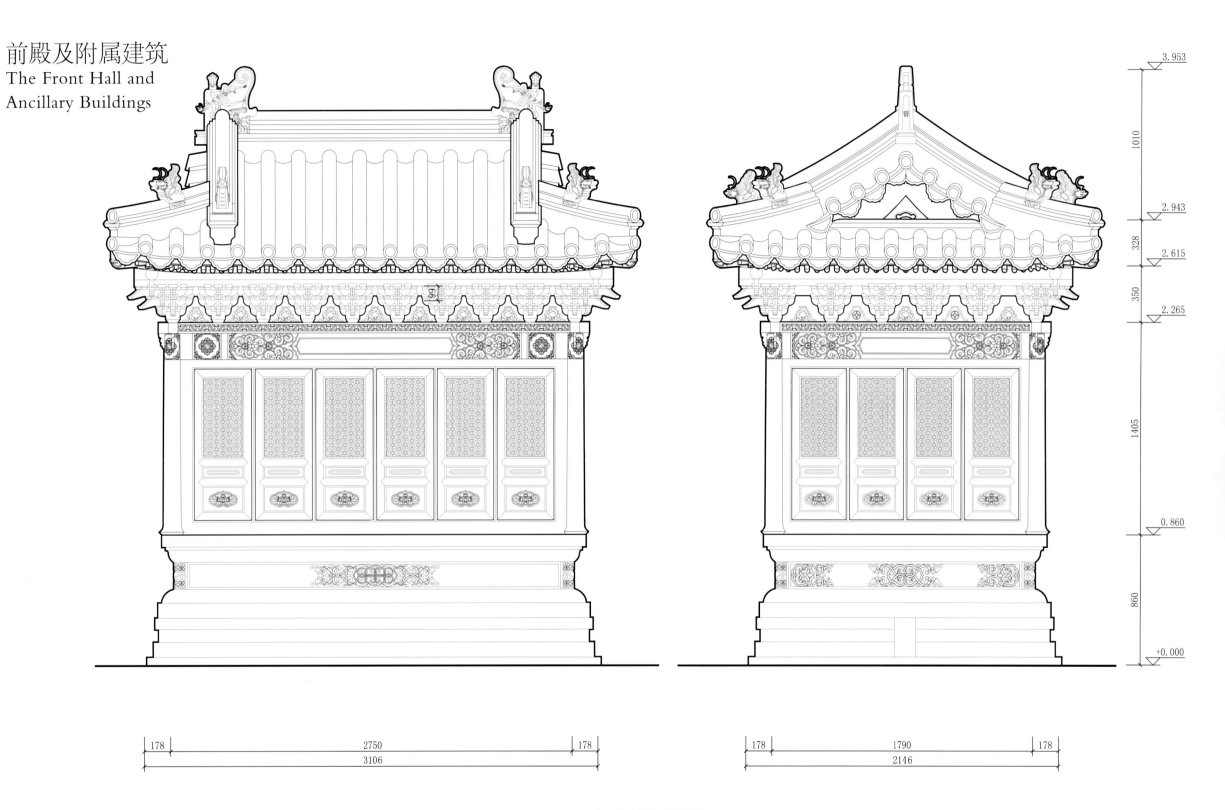

太庙焚帛炉立面图
Elevation of the Silk Burner of the Ancestral Temple

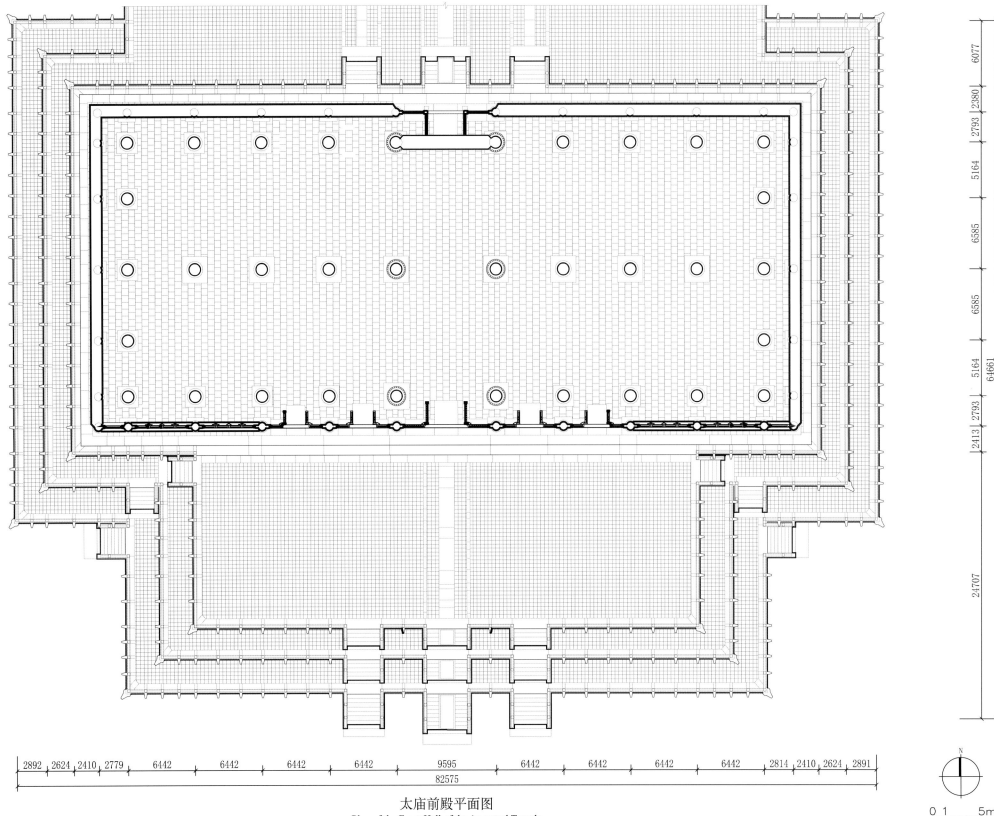

太庙前殿平面图
Plan of the Front Hall of the Ancestral Temple

2892 2624 2410 2779 6442 6442 6442 6442 9595 6442 6442 6442 6442 2814 2410 2624 2891

82575

6077 2380 2793 5164 6585 6585 5164 2793 2413

64661

24707

0 1 5m

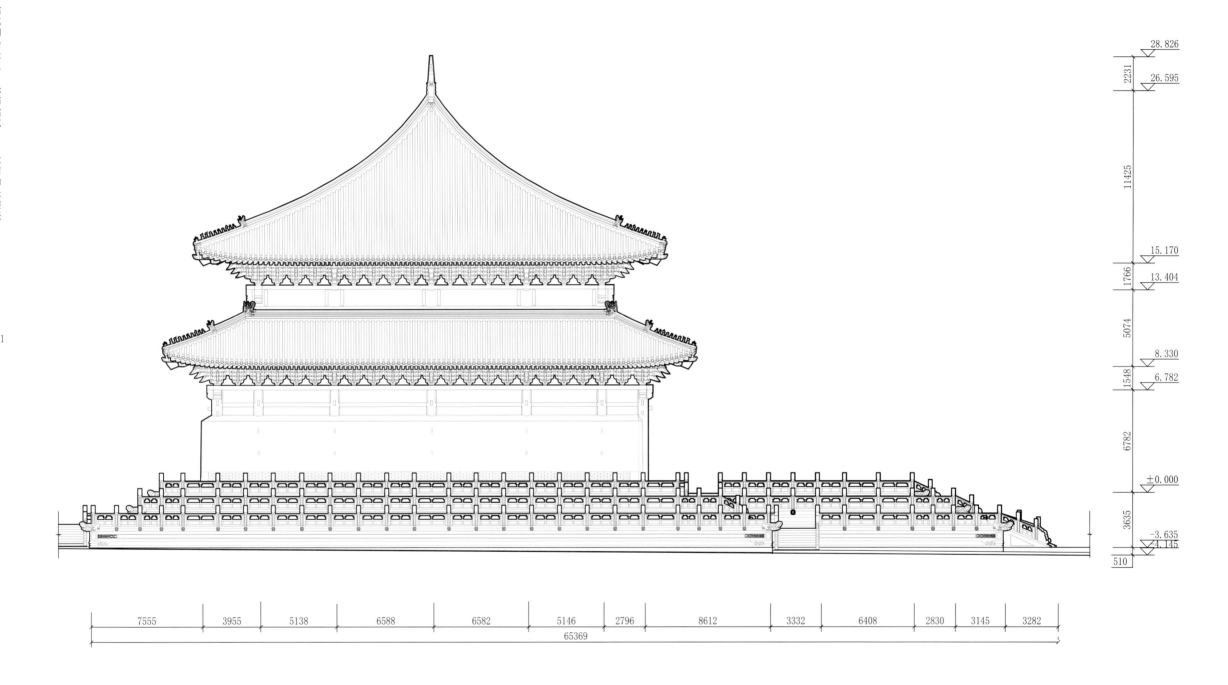

28.826

26.595

2231

11425

15.170

13.404

1766

5074

8.330

6.782

1548

6782

±0.000

3635

-3.635

-4.145

510

7555　3955　5138　6588　6582　5146　2796　8612　3332　6408　2830　3145　3282

65369

太庙前殿西立面图

West elevation of the Front Hall of the Ancestral Temple

0 1　　　5m

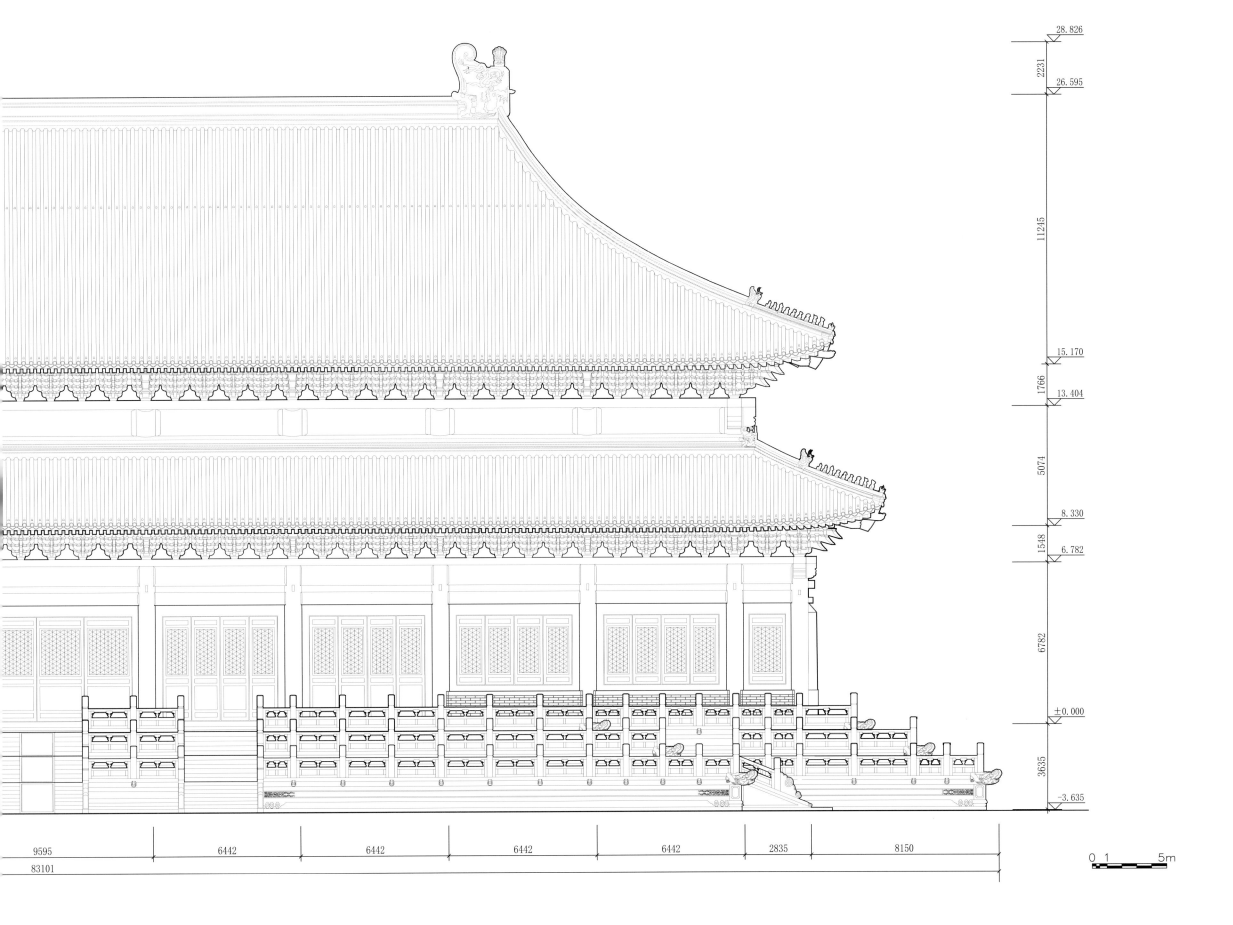

28.826

2231

26.595

11245

15.170

1766

13.404

5074

8.330

1548

6.782

6782

±0.000

3635

-3.635

9595    6442    6442    6442    6442    2835    8150

83101

0  1        5m

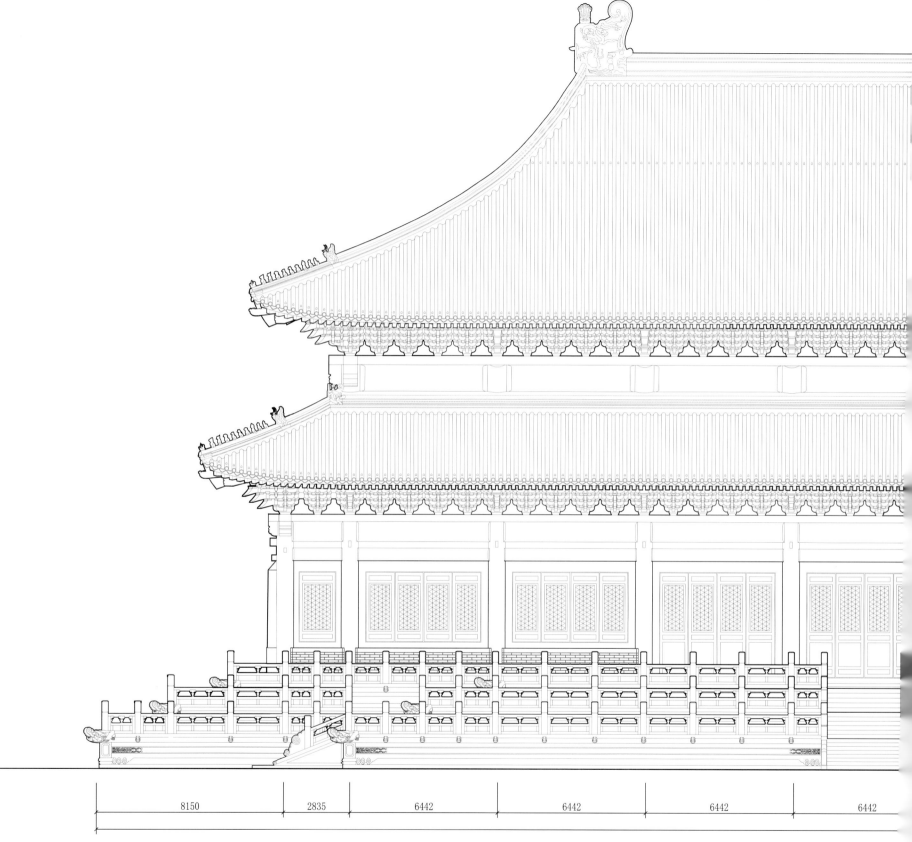

8150    2835    6442    6442    6442    6442

太庙前殿南立面图
South elevation of the Front Hall of the Ancestral Temple

28.826

2231

26.595

11245

15.170

1766

13.404

5074

8.330

1548

6.782

6782

±0.000

3635

-3.635

9595   6442   6442   6442   6442   2835   8150

83101

0  1        5m

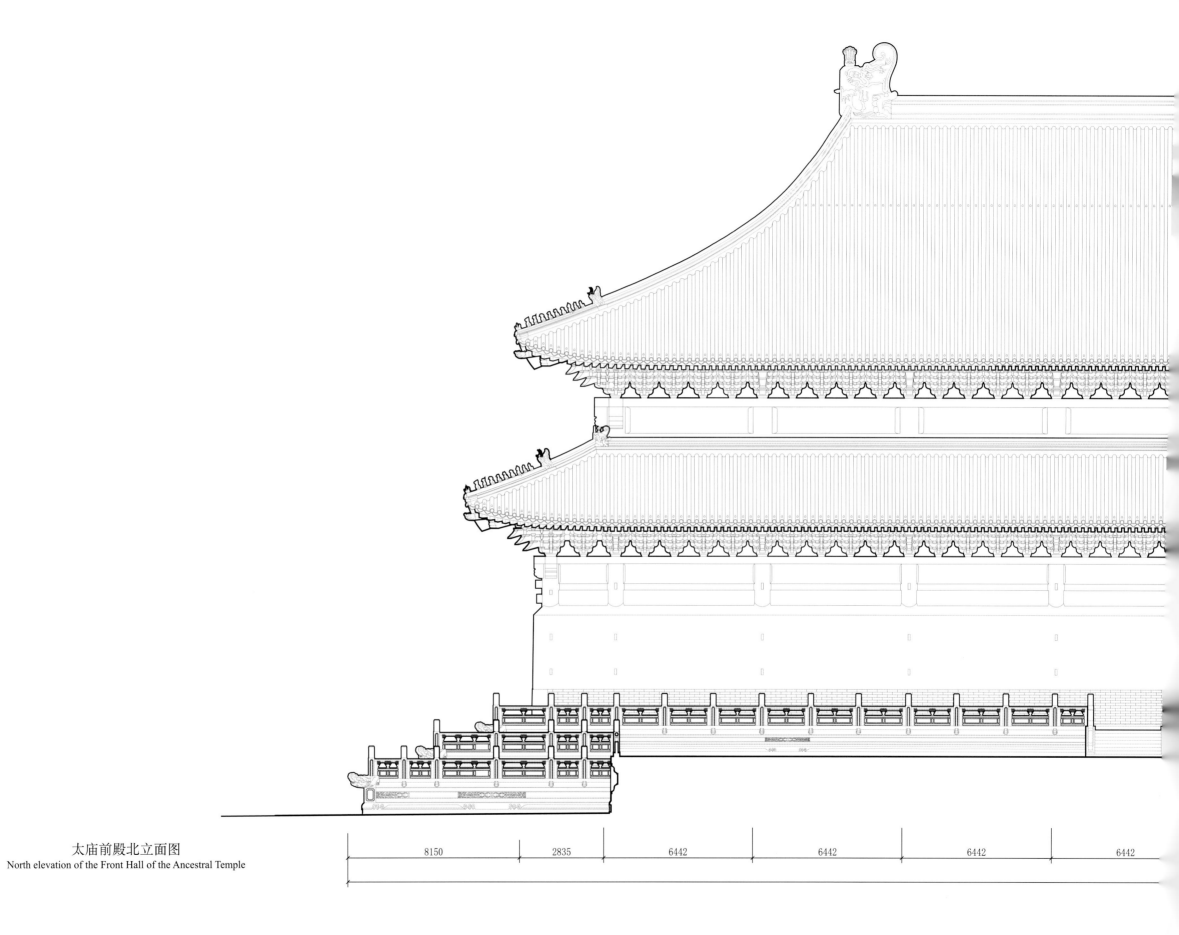

太庙前殿北立面图
North elevation of the Front Hall of the Ancestral Temple

| 8150 | 2835 | 6442 | 6442 | 6442 | 6442 |

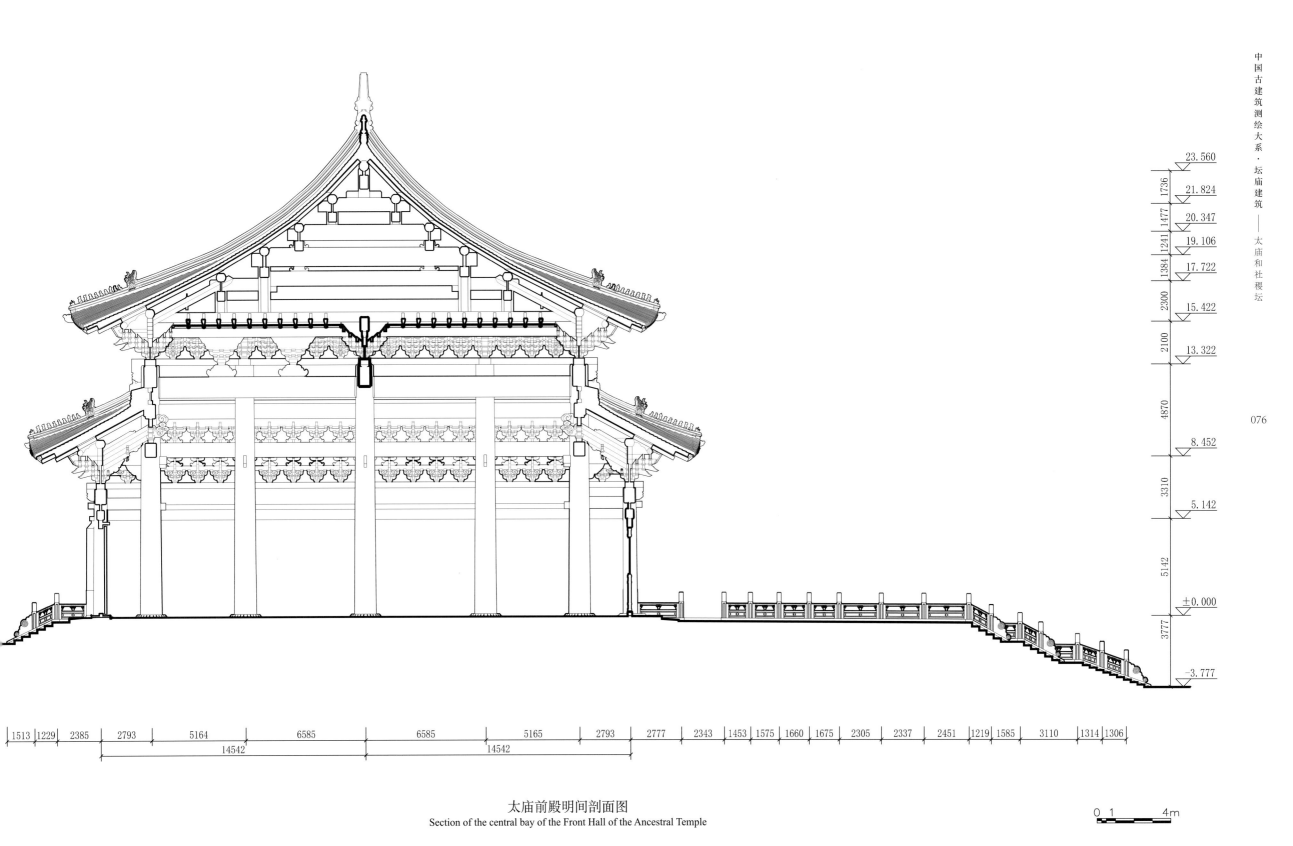

23.560

1736

21.824

1477

20.347

1241

19.106

1384

17.722

2300

15.422

2100

13.322

4870

8.452

3310

5.142

5142

±0.000

3777

-3.777

| 1513 | 1229 | 2385 | 2793 | 5164 | 6585 | 6585 | 5165 | 2793 | 2777 | 2343 | 1453 | 1575 | 1660 | 1675 | 2305 | 2337 | 2451 | 1219 | 1585 | 3110 | 1314 | 1306 |

14542　　　　14542

太庙前殿明间剖面图
Section of the central bay of the Front Hall of the Ancestral Temple

0　1　　　4m

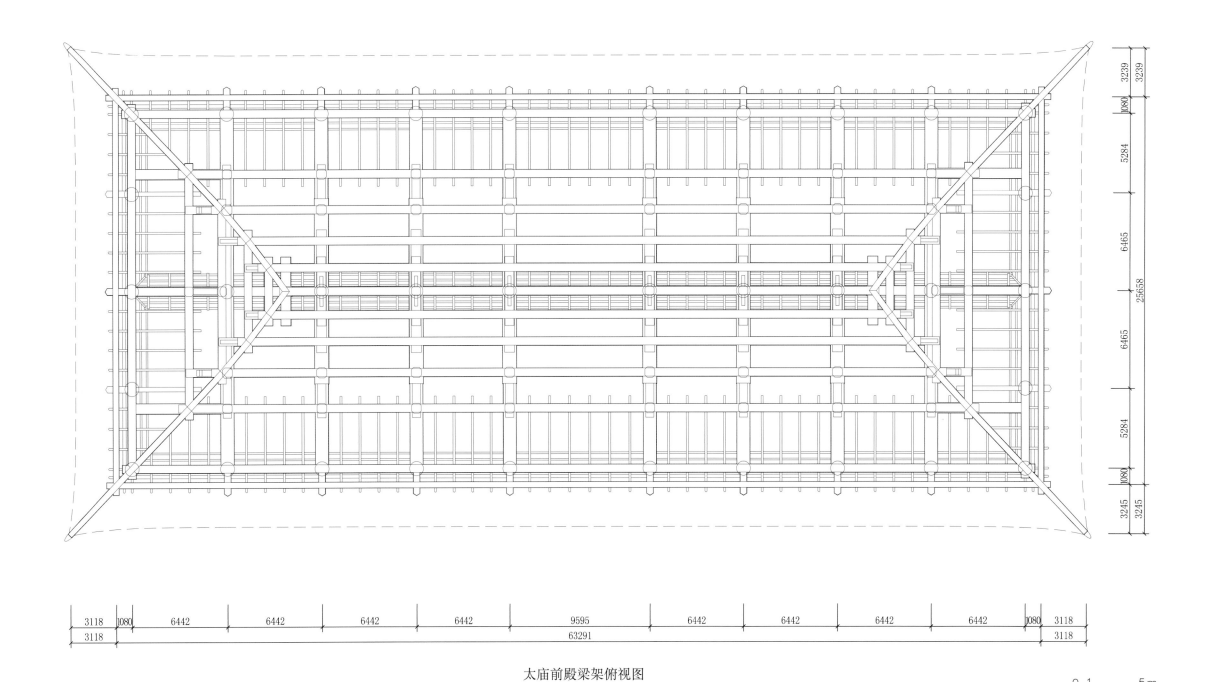

3239
1080
3239
5284
6465
25658
6465
5284
1080
3245
3245

3118
1080
6442
6442
6442
6442
9595
6442
6442
6442
6442
1080
3118
3118
63291
3118

太庙前殿梁架俯视图
Top view of the beam frame of the Front Hall of the Ancestral Temple

0 1 5m

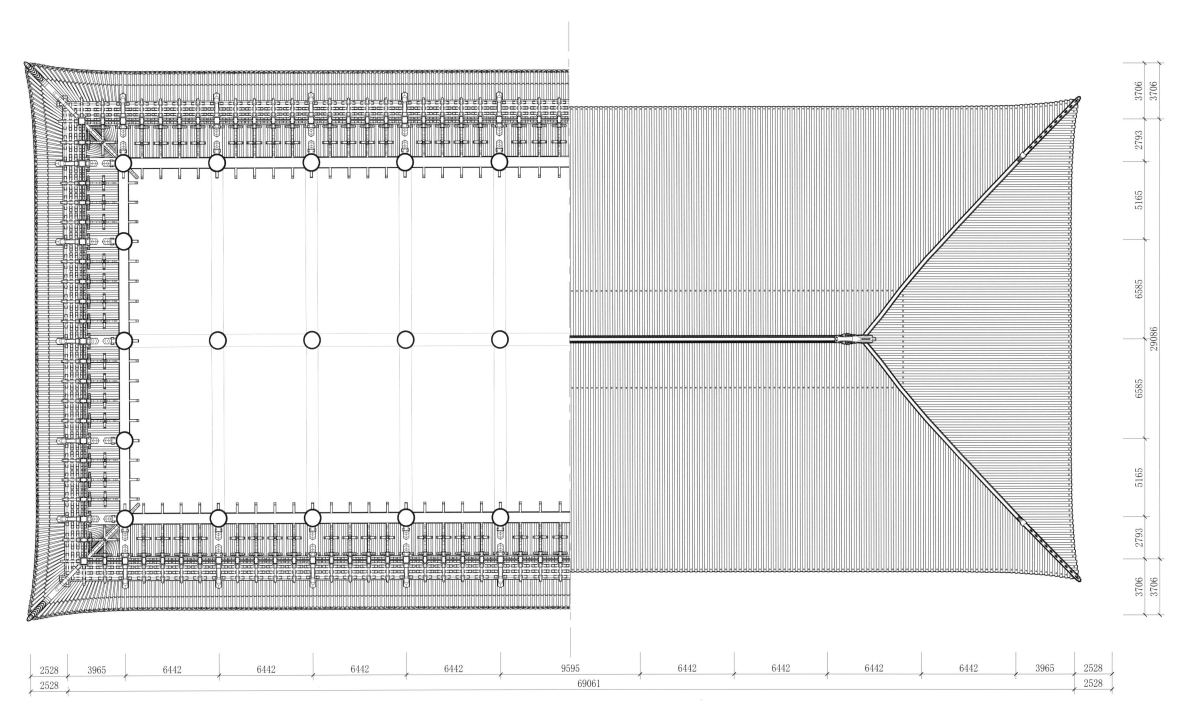

太庙前殿下檐仰视图及屋顶平面图
Bottom view of lower eaves and the roof plan of the Front hall of the Ancestral Temple

0 1                5m

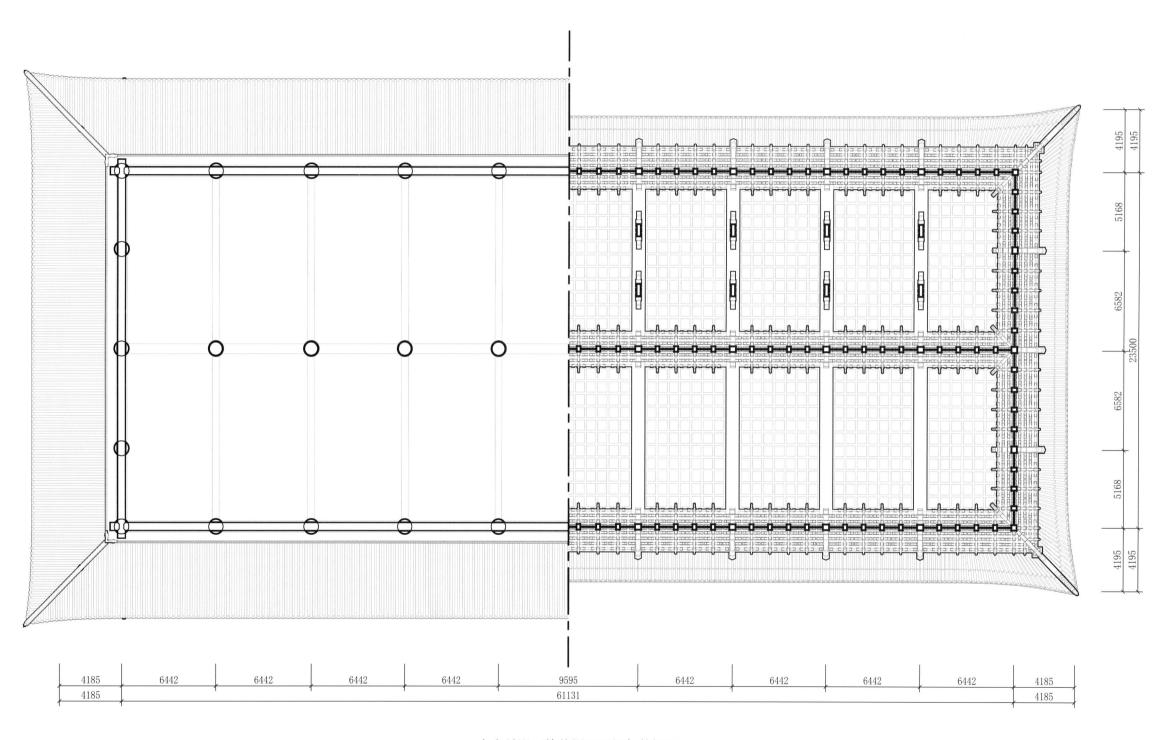

4195
4195
5168
6582
23500
6582
5168
4195
4195

4185 | 6442 | 6442 | 6442 | 6442 | 9595 | 6442 | 6442 | 6442 | 6442 | 4185
4185
61131
4185

太庙前殿下檐俯视图及梁架仰视图
Top view of lower eaves and the beam frame of the Front hall of the Ancestral Temple

0 1　　　5m

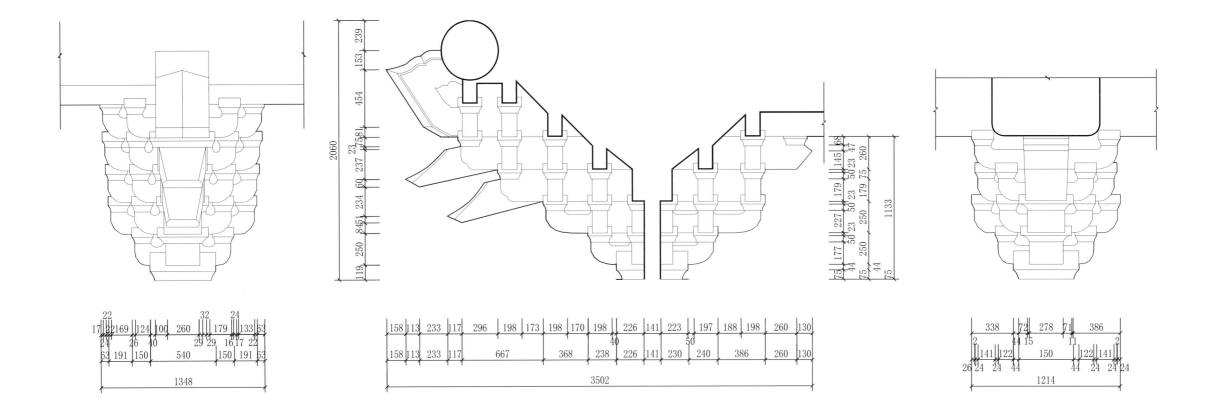

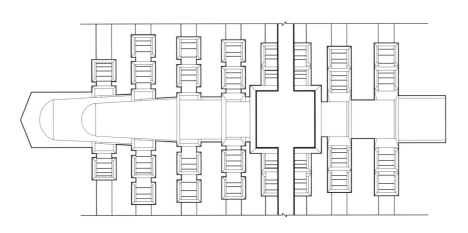

太庙前殿上檐柱头科斗栱大样图

Detail of the connecting bracket sets on the outer columns of upper eaves of the Front Hall of the Ancestral Temple

0    0.3    0.6m

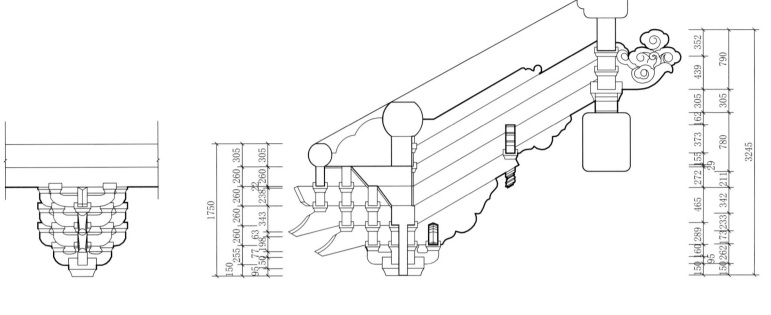

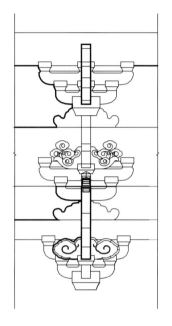

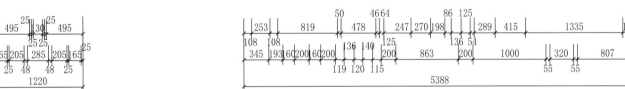

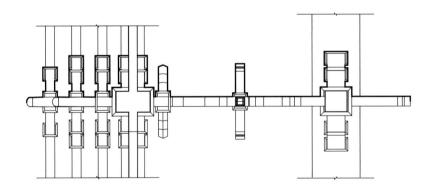

太庙前殿下檐平身科镏金斗栱大样图

Detail of the cantilevered bracket sets between the columns of lower eaves of the Front Hall of the Ancestral Temple

0　0.5　1m

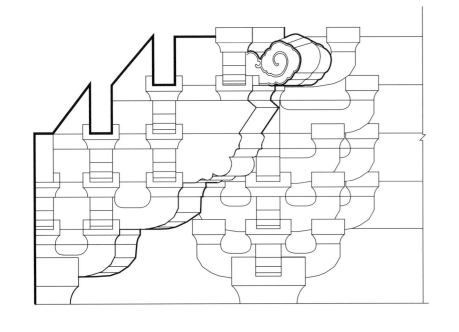

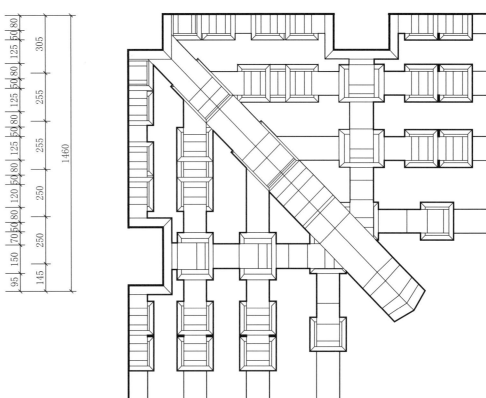

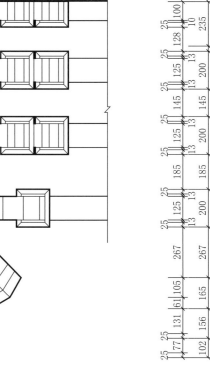

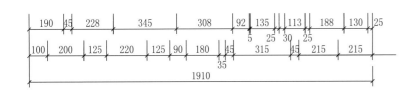

太庙前殿内檐角科斗栱大样图

Detail of the connecting bracket sets on the inner corner columns of the Front Hall of the Ancestral Temple

0    0.3    0.6m

大庙前殿前丹陛大样图之一

Front palatial steps detail drawing I of the Front Hall of the Ancestral Temple

0  0.1        0.3m

太庙前殿前丹陛大样图之二

Front palatial steps detail drawing II of the Front Hall of the Ancestral Temple

太庙前殿前殿前丹陛大样图之三

Front palatial steps detail drawing III of the Front Hall of the Ancestral Temple

太庙大殿殿后丹陛

Rear palatial steps detail drawing of the Front Hall of the Ancestral Temple

太庙前殿望柱头展开图之一
Expanded view I of the balustrade of the Front Hall of the Ancestral Temple

太庙前殿望柱头展开图之二
Expanded view II of the balustrade of the Front Hall of the Ancestral Temple

430

800

太庙前殿望柱头展开图之三
Expanded view III of the balustrade of the Front Hall of the Ancestral Temple

420

780

太庙前殿望柱头展开图之四
Expanded view IV of the balustrade of the Front Hall of the Ancestral Temple

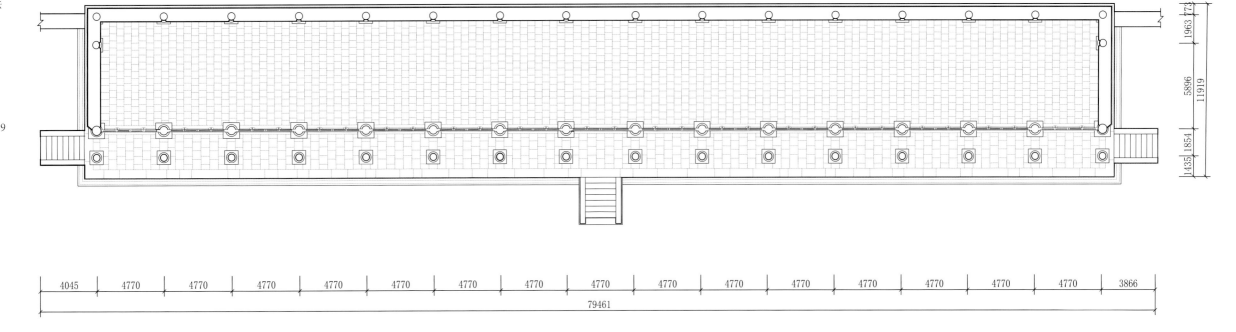

4045 4770 4770 4770 4770 4770 4770 4770 4770 4770 4770 4770 4770 4770 4770 3866

79461

太庙前殿配殿平面图
Plan of the side hall in front of the Ancestral Temple

0　1　5m

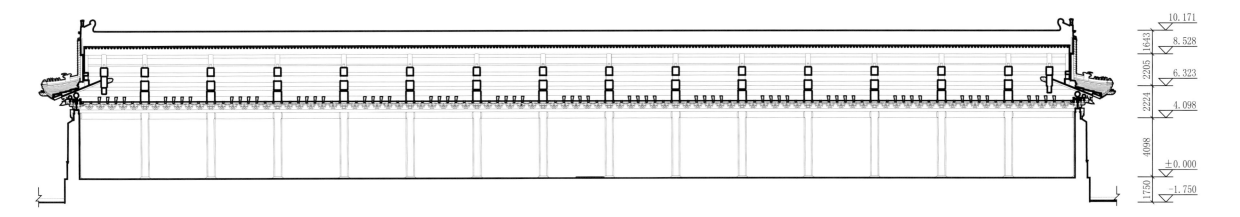

10. 171
8. 528
6. 323
4. 098
±0. 000
-1. 750

1643
2205
2224
4098
1750

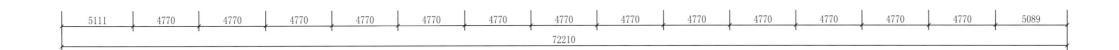

5111　4770　4770　4770　4770　4770　4770　4770　4770　4770　4770　4770　4770　4770　5089

72210

太庙前殿配殿纵剖面图
Longitudinal section of the side hall in front of the Ancestral Temple

0 1　　　5m

091

11.254

6114

4.840

4840

0.000

1750

−1.750

| 4045 | 4770 | 4770 | 4770 | 4770 | 4770 | 4770 | 4770 | 4770 | 4770 | 4770 | 4770 | 4770 | 4770 | 4770 | 4770 | 3866 |

79461

太庙前殿配殿正立面图
Front elevation of the side hall in front of the Ancestral Temple

0　1　2m

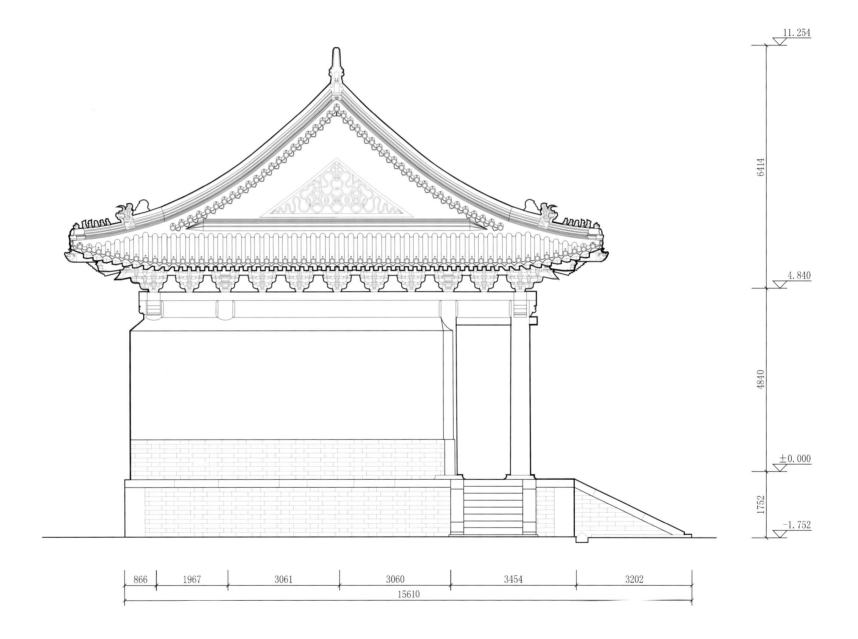

11.254

6114

4.840

4840

±0.000

1752

−1.752

866　1967　3061　3060　3454　3202

15610

太庙前殿配殿侧立面图
Side elevation of the side hall in front of the Ancestral Temple

0　1　3m

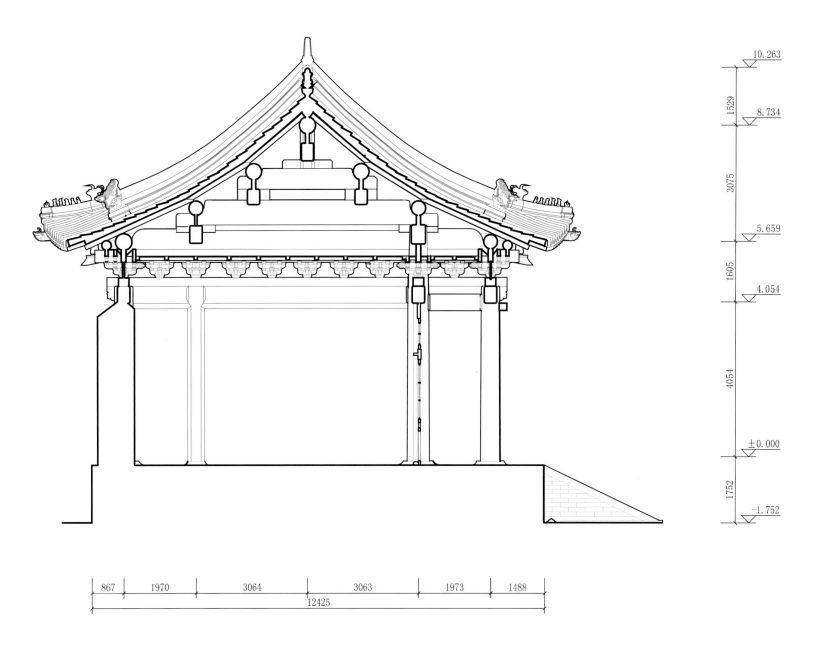

10.263

1529

8.734

3075

5.659

1605

4.054

4054

±0.000

1752

-1.752

867　1970　3064　3063　1973　1488

12425

太庙前殿配殿明间剖面图
Section of the central bay of the side hall in front of the Ancestral Temple

0　1　3m

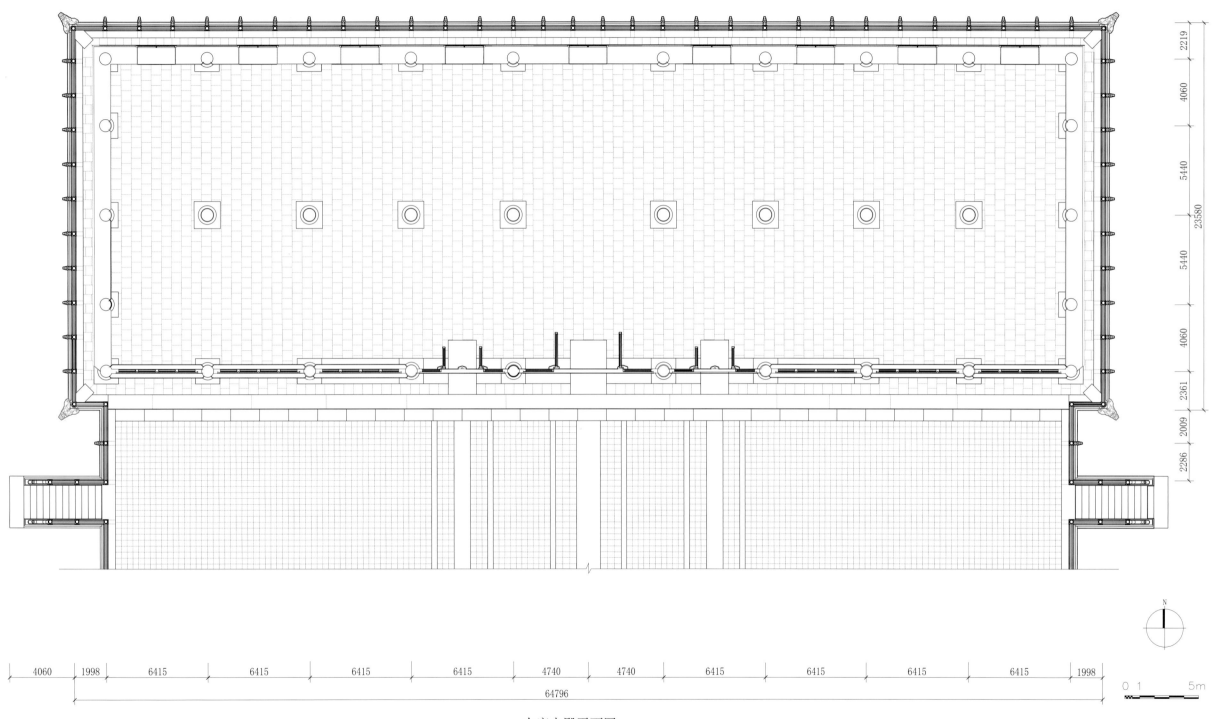

太庙中殿平面图
Plan of the central hall of the Ancestral Temple

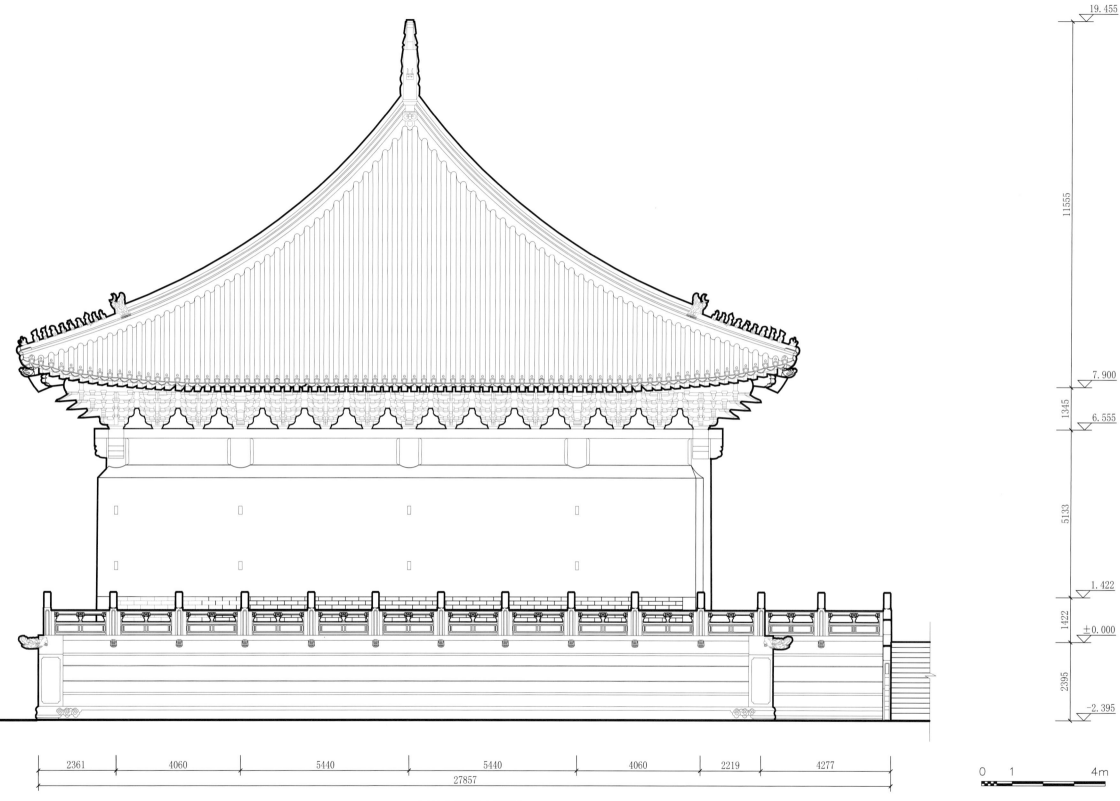

19.455

11555

7.900

1345

6.555

5133

1.422

1422

±0.000

2395

-2.395

2361　4060　5440　5440　4060　2219　4277

27857

0　1　　　4m

太庙中殿西立面图
West elevation of the central hall of the Ancestral Temple

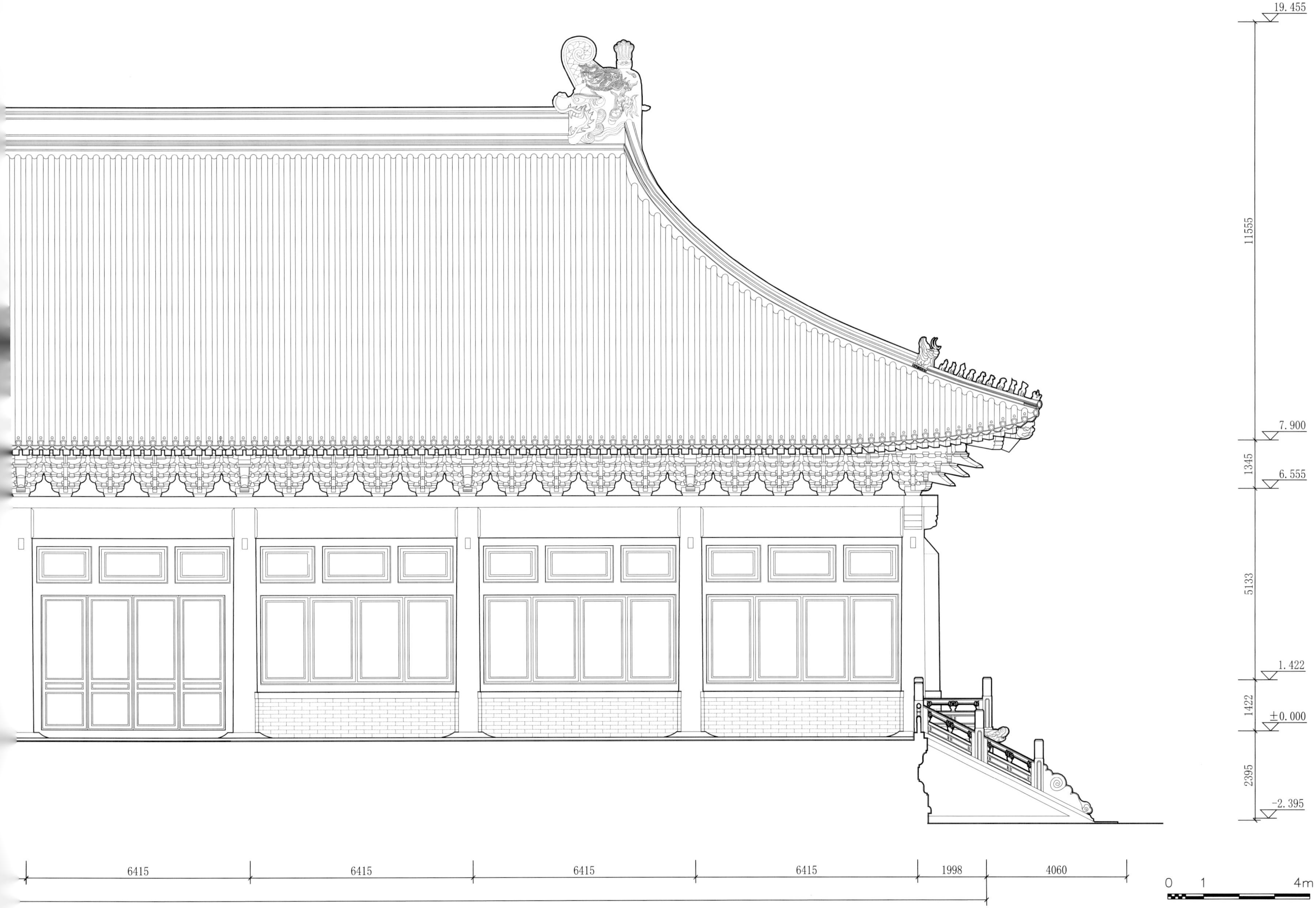

19.455

11555

7.900

1345

6.555

5133

1.422

1422

±0.000

2395

-2.395

6415  6415  6415  6415  1998  4060

0   1   4m

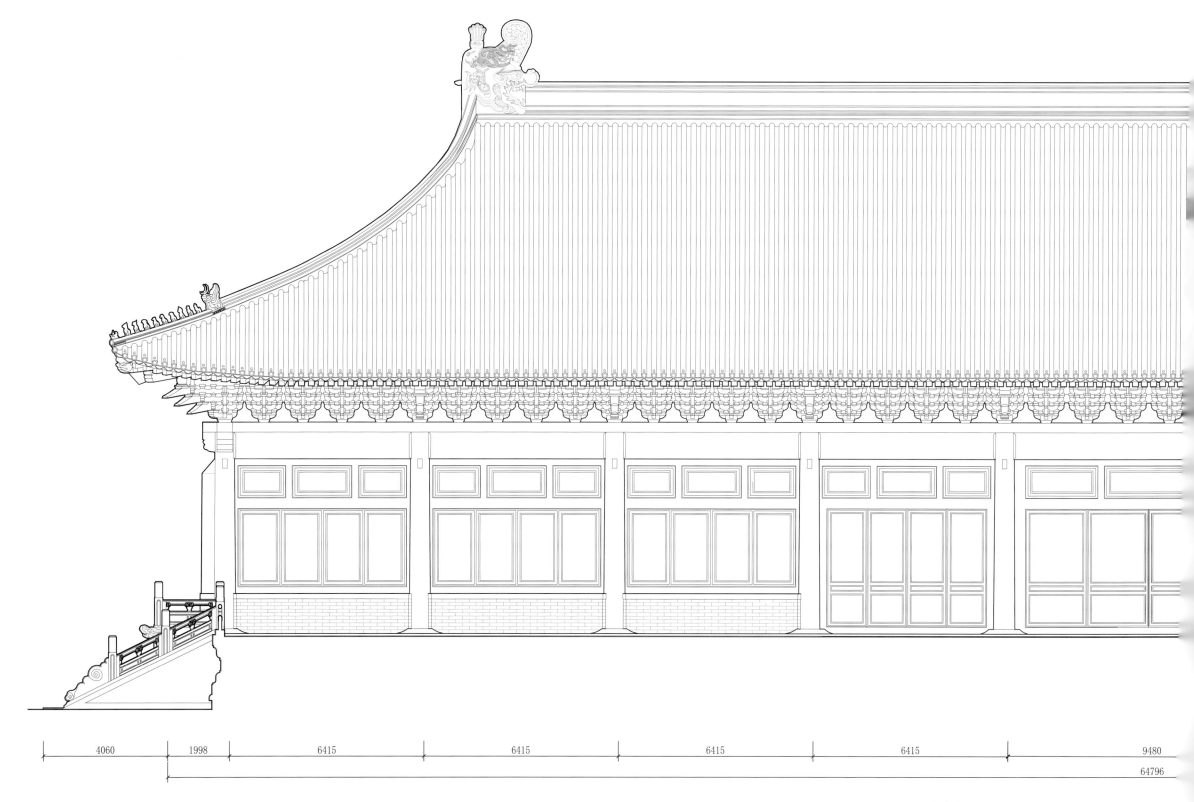

4060  1998  6415  6415  6415  6415  9480

64796

太庙中殿南立面图
South elevation of the central hall of the Ancestral Temple

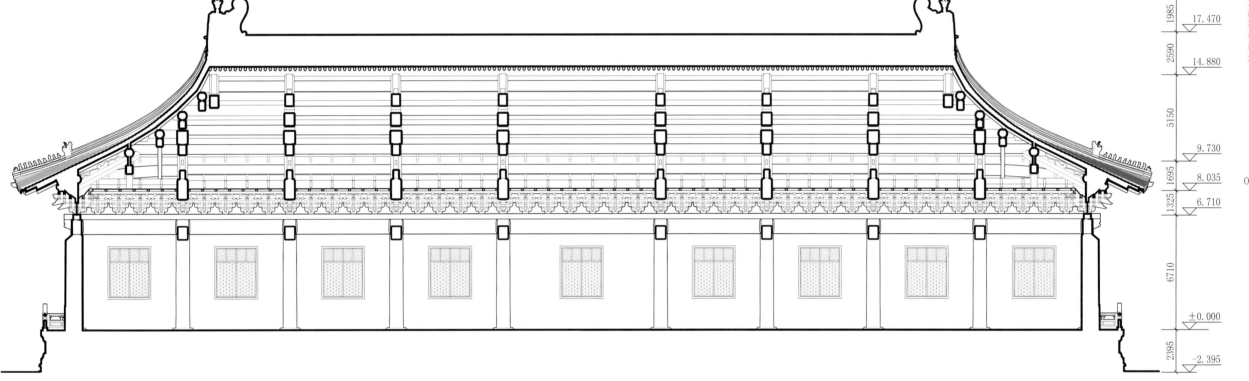

19.455

1985

17.470

2590

14.880

5150

9.730

1695

8.035

1325

6.710

6710

±0.000

2395

−2.395

| 1890 | 6722 | 6430 | 6430 | 6430 | 9480 | 6430 | 6430 | 6430 | 6430 | 1890 |

64992

太庙中殿纵剖面图
Longitudinal section of the central hall of the Ancestral Temple

0 1    4m

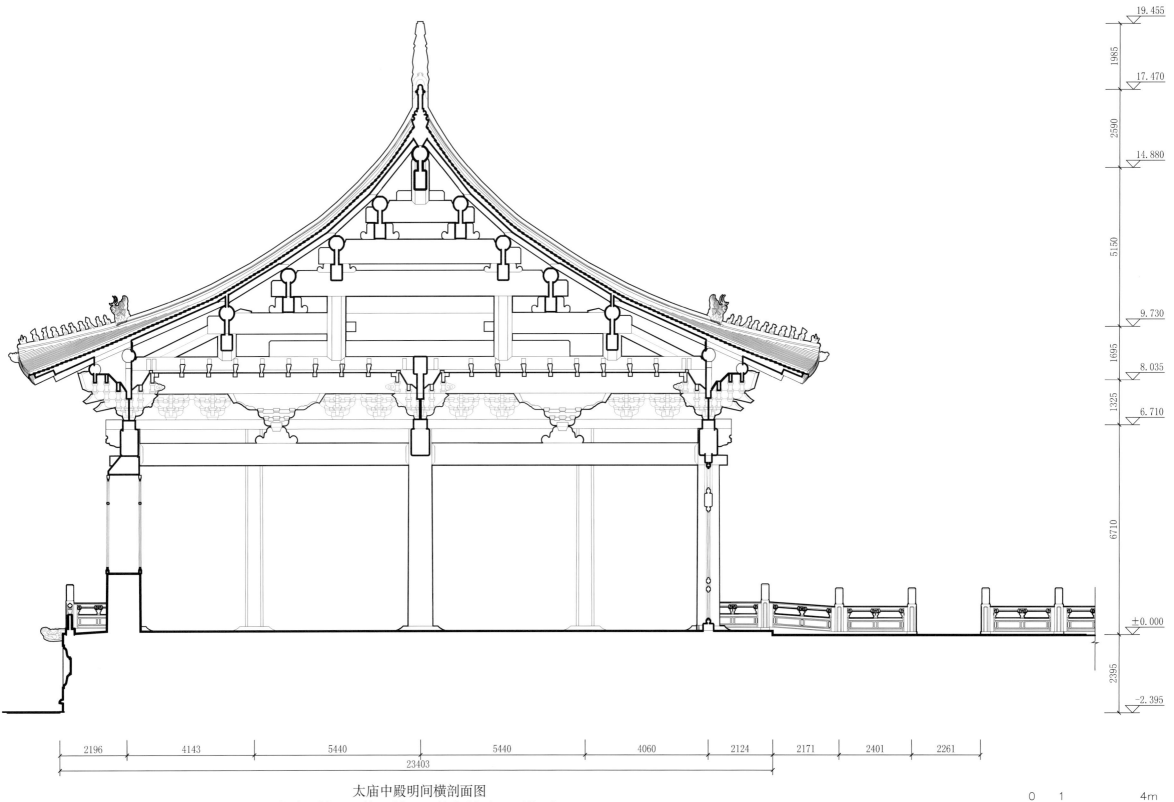

19. 455

17. 470

14. 880

9. 730

8. 035

6. 710

±0. 000

-2. 395

1985

2590

5150

1695

1325

6710

2395

2196    4143    5440    5440    4060    2124    2171    2401    2261

23403

太庙中殿明间横剖面图
Section of the central bay of the central hall of the Ancestral Temple

0    1    4m

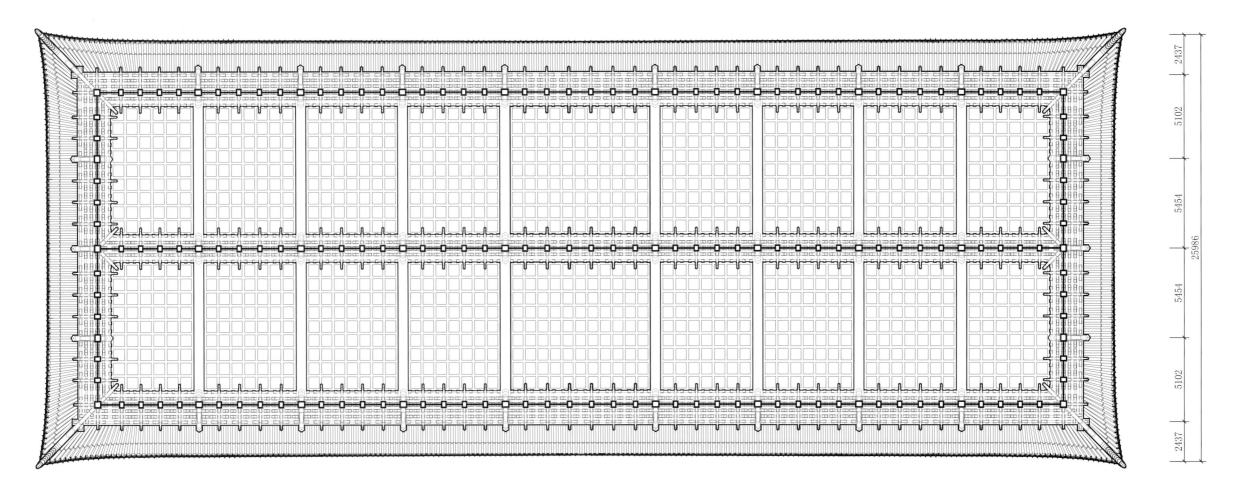

太庙中殿天花仰视图
Bottom view of the ceiling of the central hall of the Ancestral Temple

0 1     4m

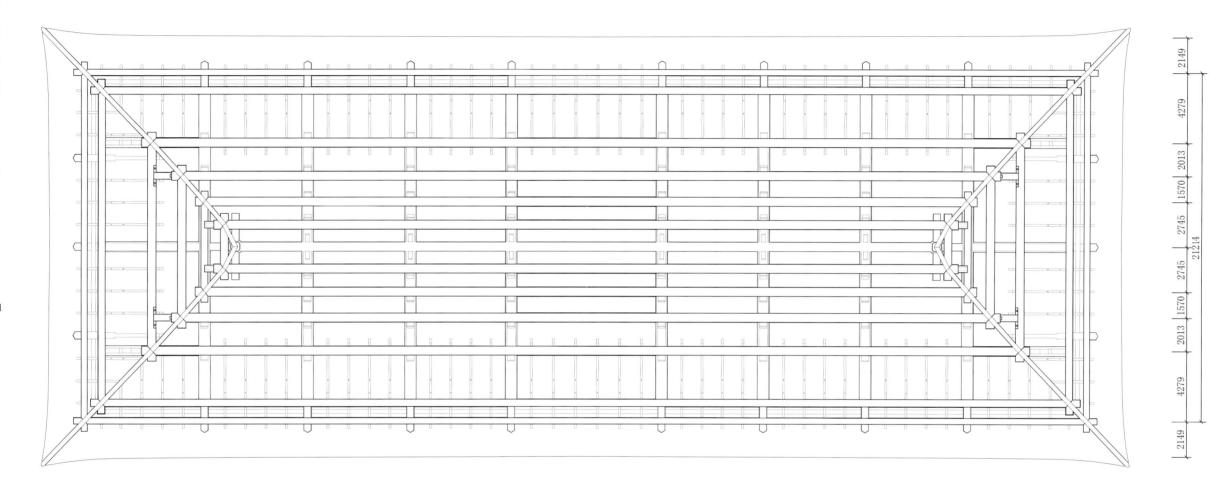

太庙中殿梁架俯视图
Top view of the beam frame of the central hall of the Ancestral Temple

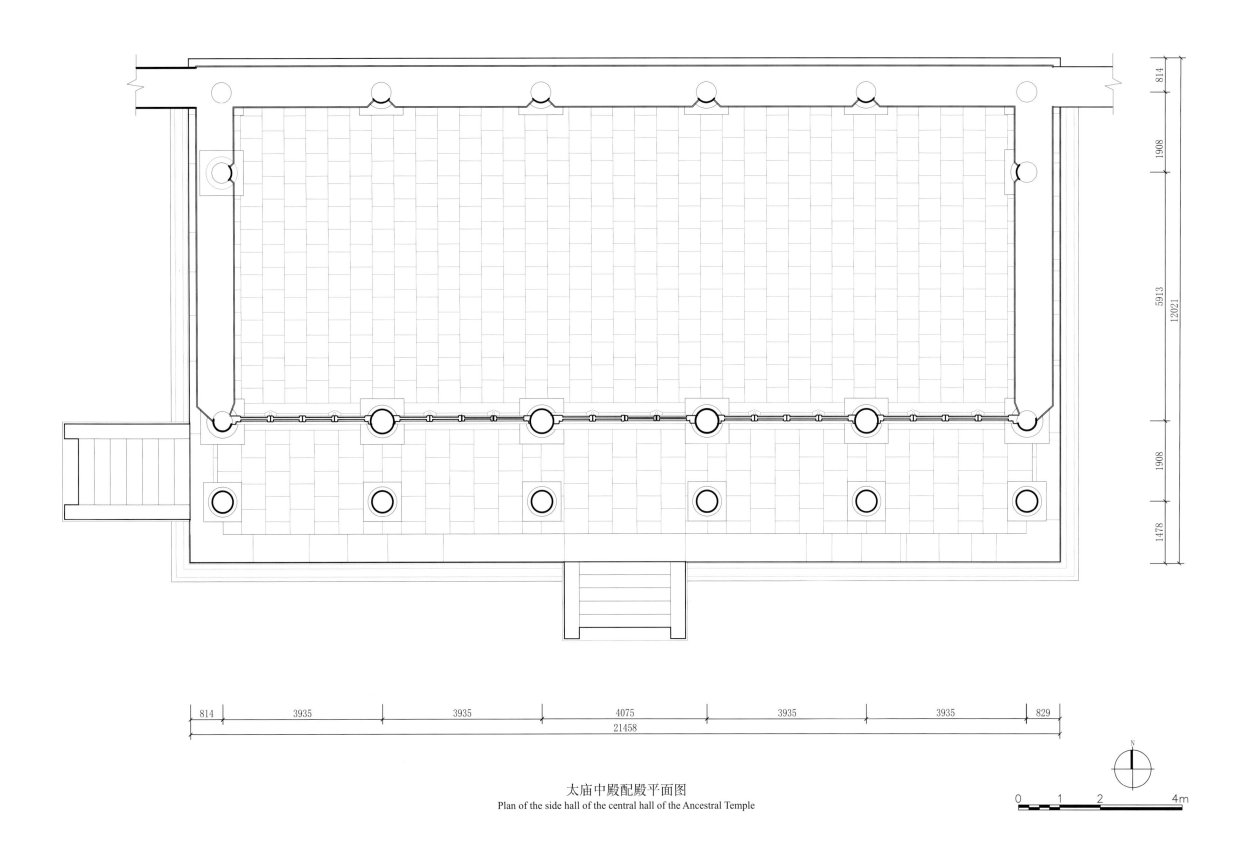

太庙中殿配殿平面图
Plan of the side hall of the central hall of the Ancestral Temple

0 1 2 4m

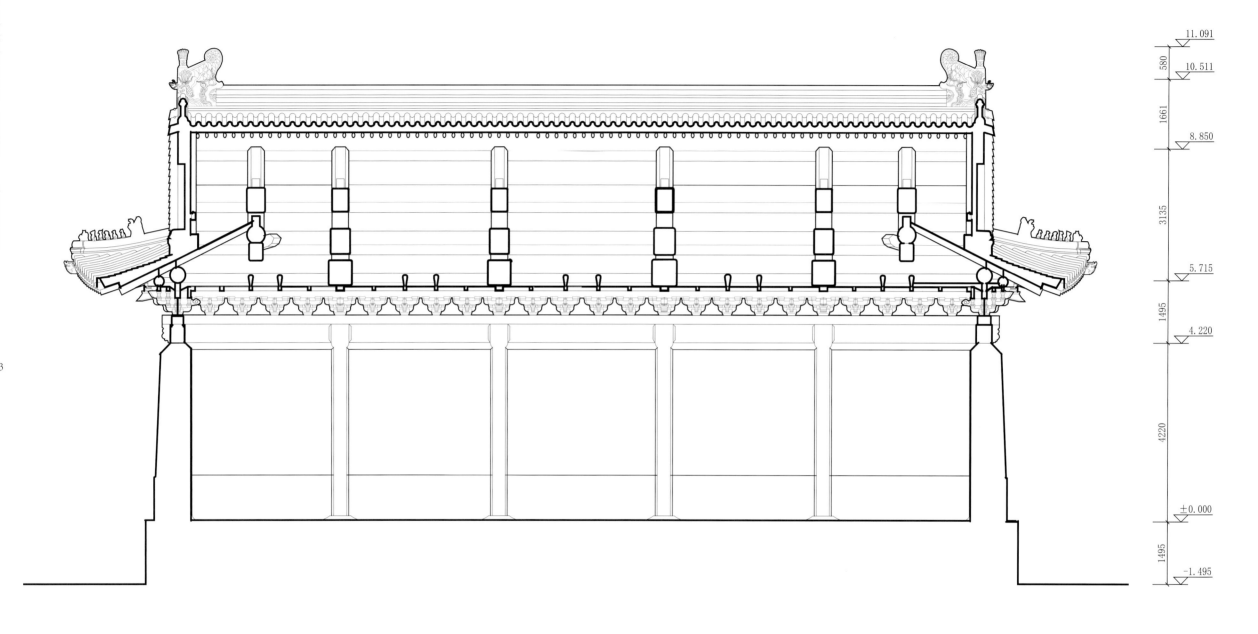

11.091
580
10.511
1661
8.850
3135
5.715
1495
4.220
4220
±0.000
1495
-1.495

814  3935  3935  4075  3935  3935  829
21458

太庙中殿配殿纵剖面
Longitudinal section of the side hall of the central hall of the Ancestral Temple

0  1  2  4m

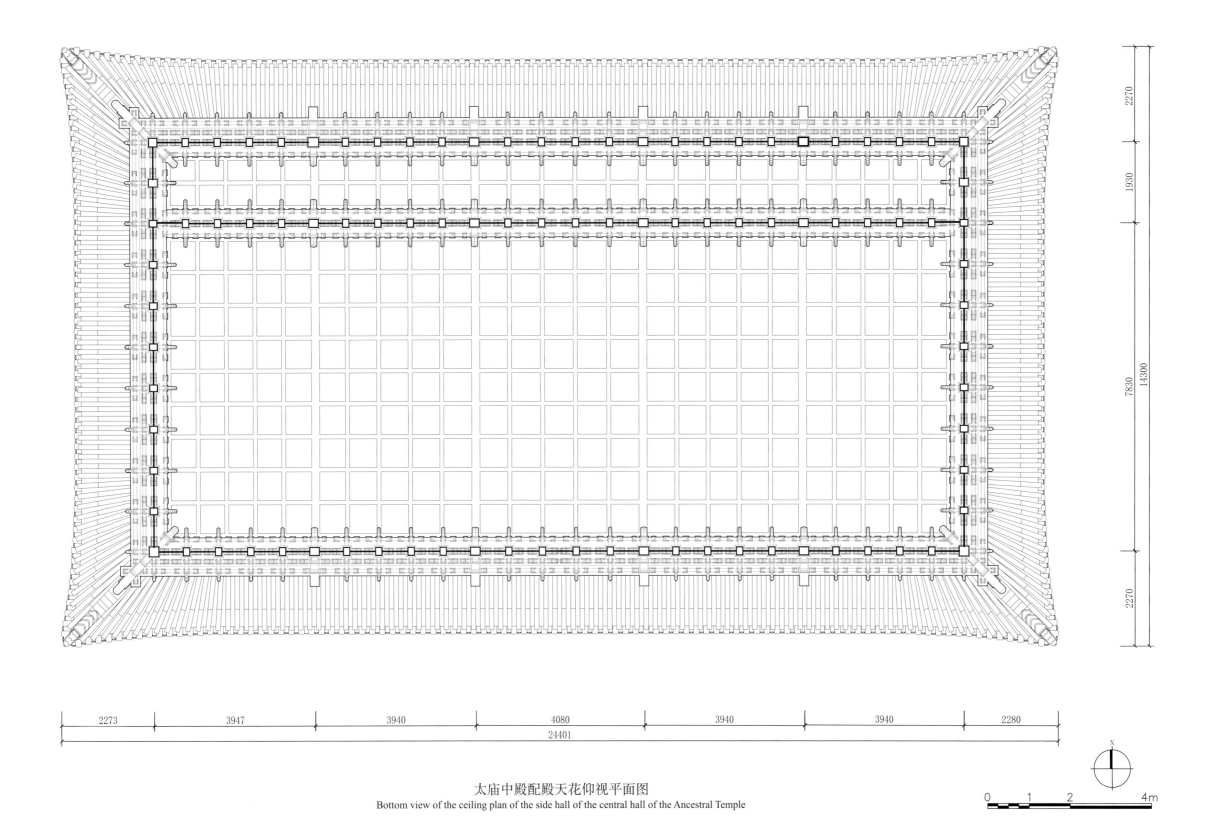

2273　3947　3940　4080　3940　3940　2280

24401

2270　1930　7830　2270

14300

太庙中殿配殿天花仰视平面图

Bottom view of the ceiling plan of the side hall of the central hall of the Ancestral Temple

0　1　2　4m

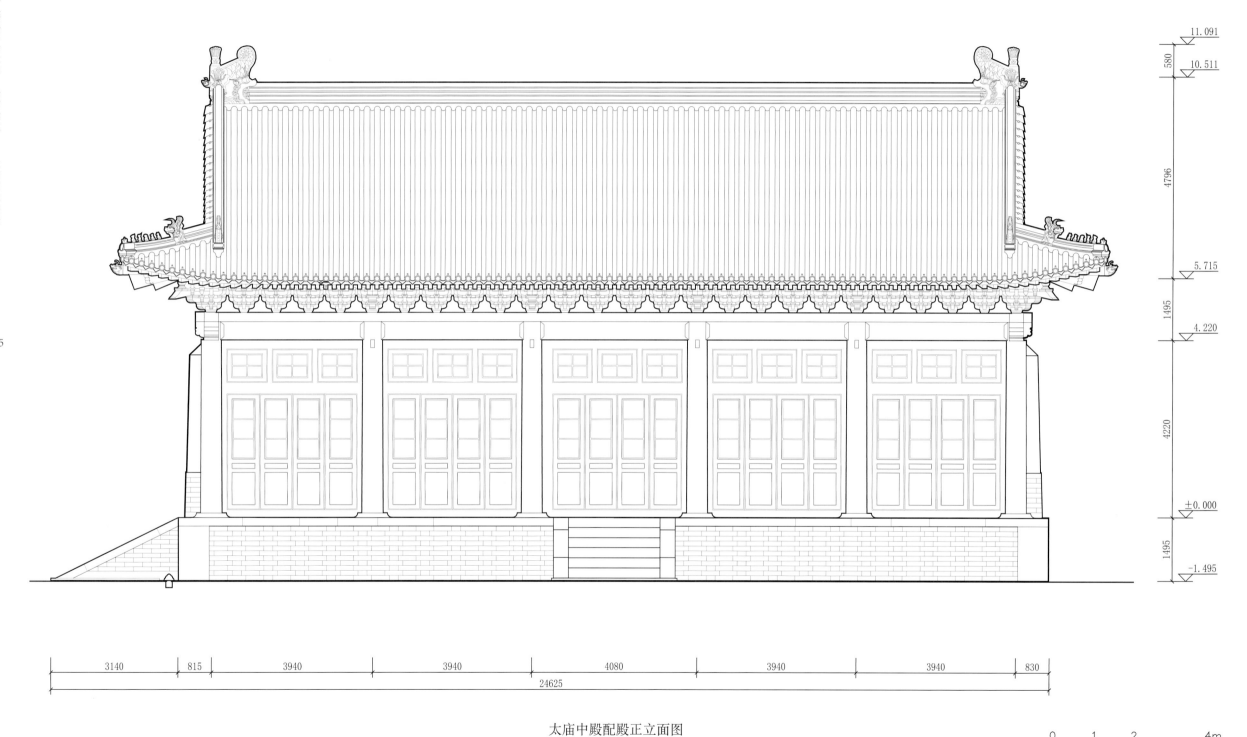

11.091
580
10.511
4796
5.715
1495
4.220
4220
±0.000
1495
-1.495

3140　815　3940　3940　4080　3940　3940　830
24625

太庙中殿配殿正立面图
Front elevation of the side hall of the central hall of the Ancestral Temple

0　1　2　4m

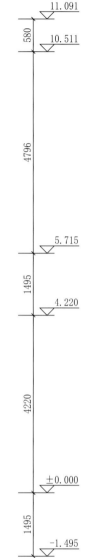

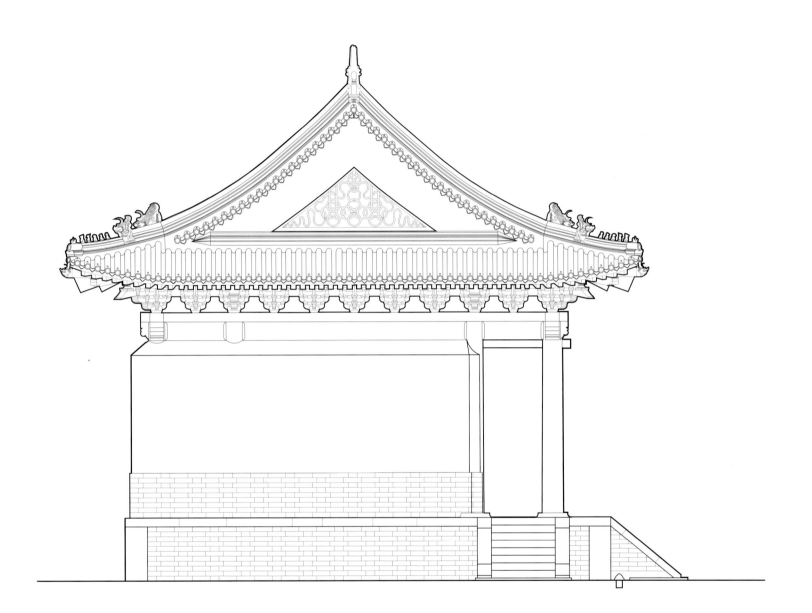

太庙中殿配殿侧立面图
Side elevation of the side hall of the central hall of the Ancestral Temple

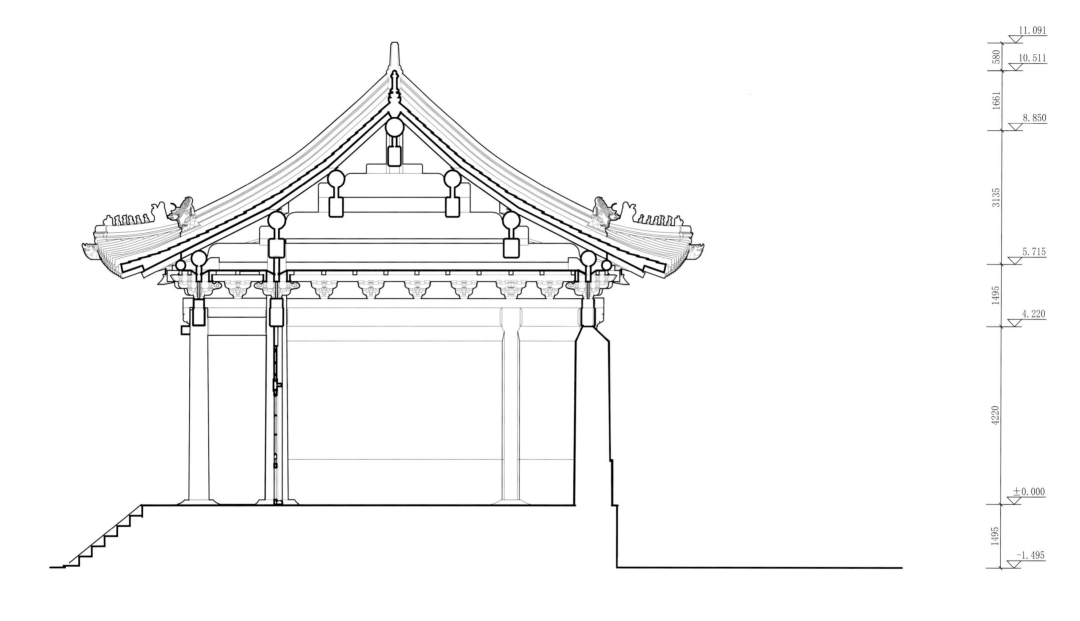

11.091
580
10.511
1661
8.850

3135

5.715
1495
4.220

4220

±0.000

1495
-1.495

| 1391 | 1910 | 5731 | 1910 | 723 |
| 11665 |

太庙中殿配殿明间剖面
Section of the central bay of the side hall of the central hall of the Ancestral Temple

0    1    2         4m

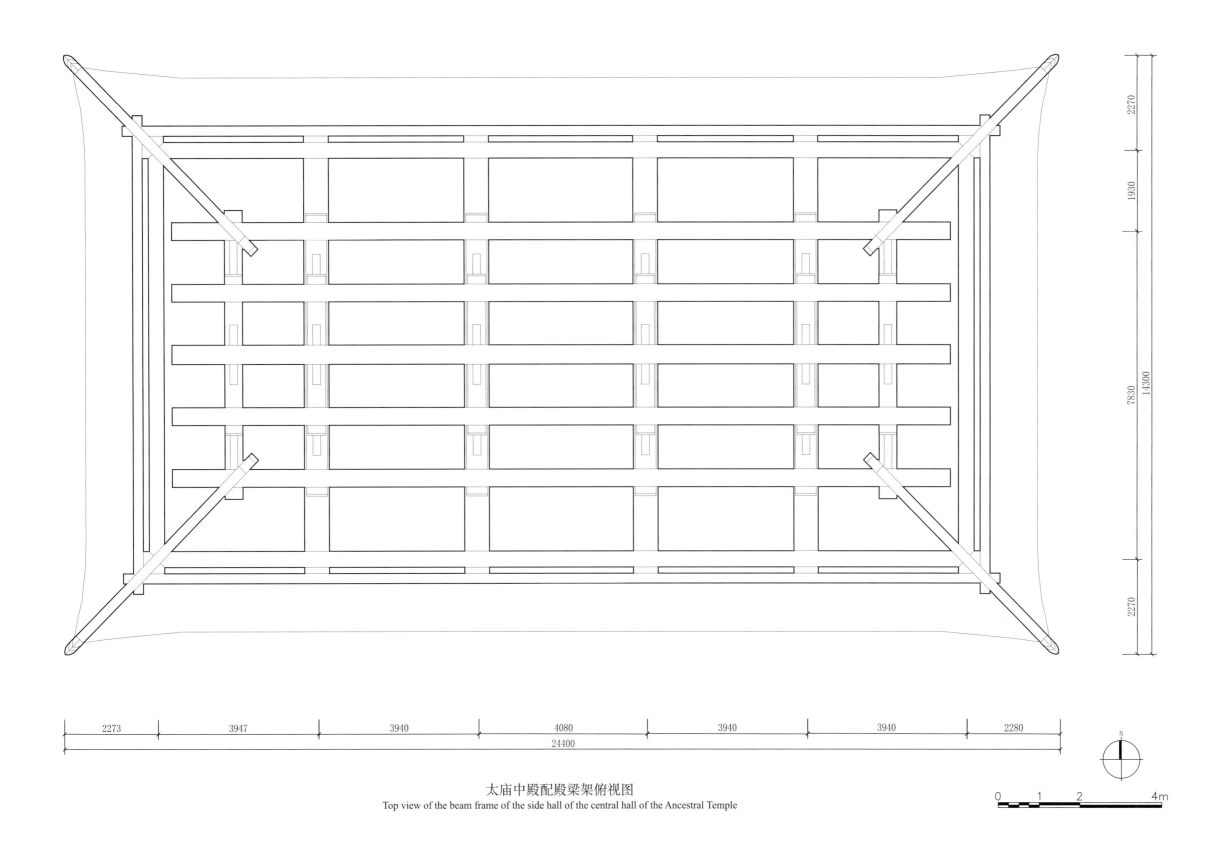

太庙中殿配殿梁架俯视图
Top view of the beam frame of the side hall of the central hall of the Ancestral Temple

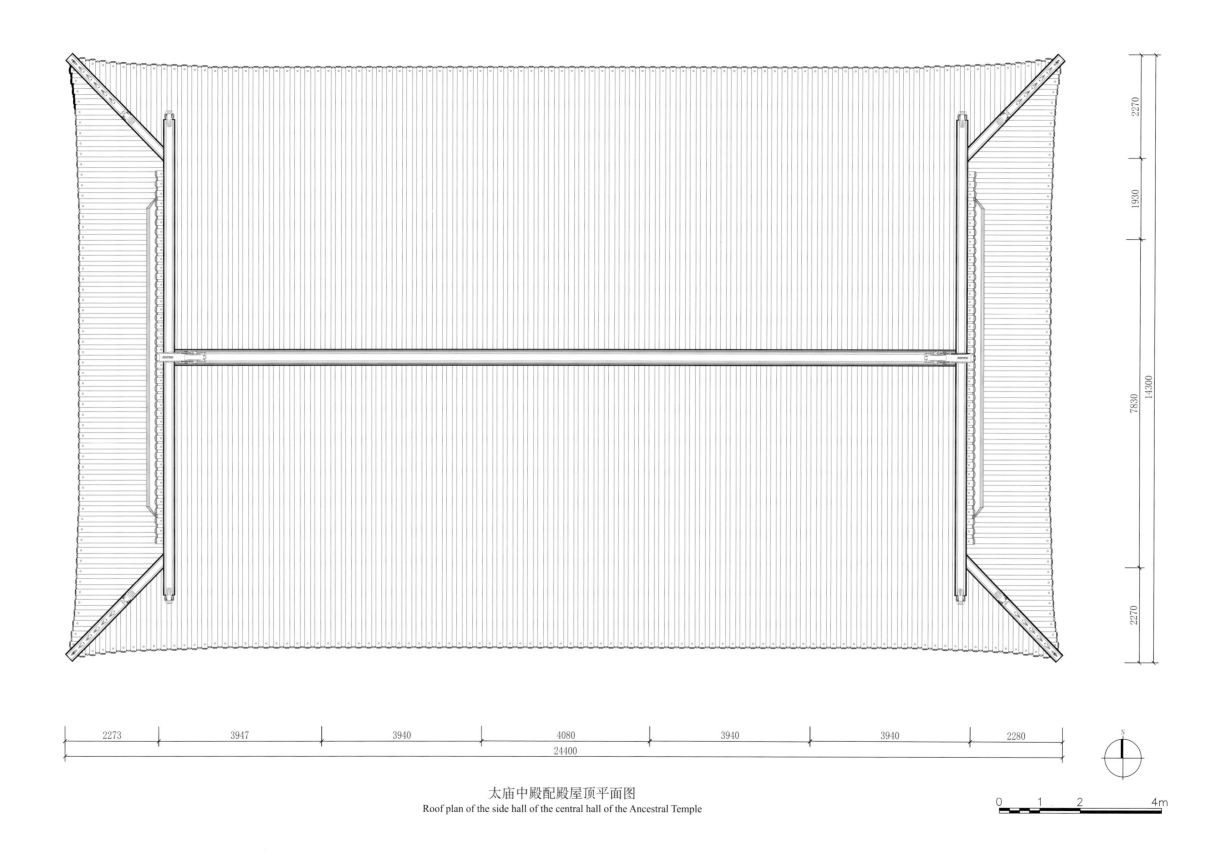

太庙中殿配殿屋顶平面图
Roof plan of the side hall of the central hall of the Ancestral Temple

2273　　3947　　3940　　4080　　3940　　3940　　2280
24400

2270
1930
7830
14300
2270

0　1　2　4m

中国古建筑测绘大系·坛庙建筑 —— 太庙和社稷坛

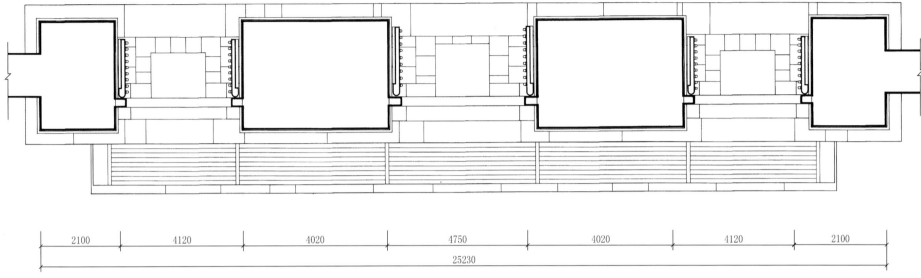

太庙二道门平面图
Plan of the Second Gate to the Ancestral Temple

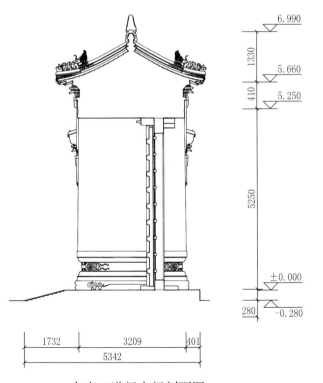

太庙二道门中门剖面图
Middle entrance section of the Second Gate to the Ancestral Temple

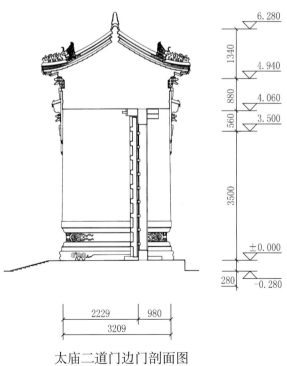

太庙二道门边门剖面图
Side entrance section of the Second Gate to the Ancestral Temple

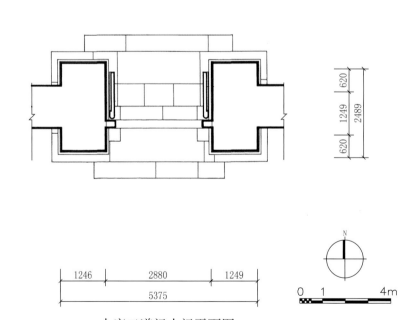

太庙二道门小门平面图
Small entrance plan of the Second Gate to the Ancestral Temple

0  1  4m

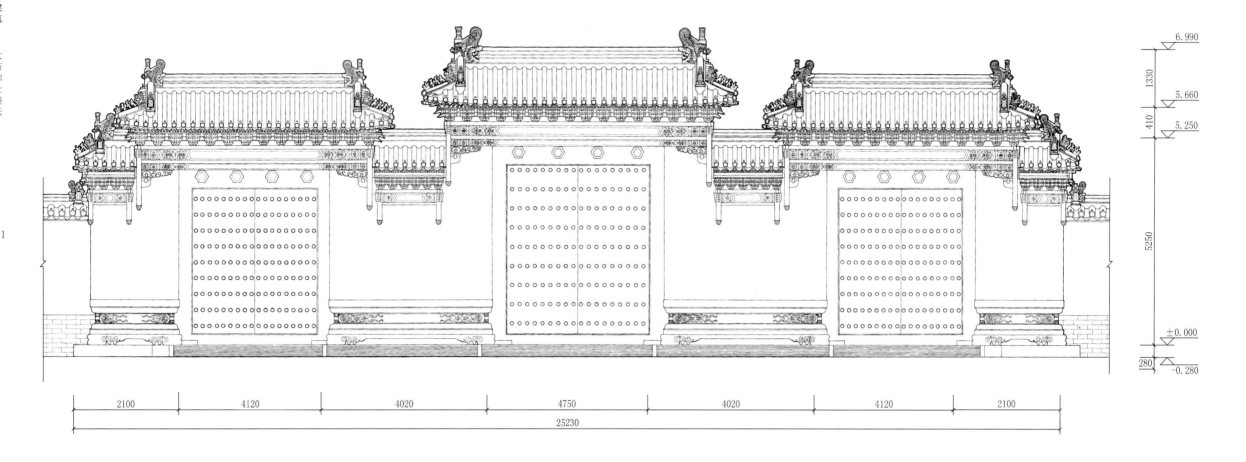

6.990

1330

5.660

410

5.250

5250

±0.000

280

-0.280

2100　4120　4020　4750　4020　4120　2100

25230

太庙二道门南立面图
South elevation of the Second Gate to the Ancestral Temple

0　1　3m

6.280

1340

4.940

880

4.060

560

3.500

3500

±0.000

280

-0.280

2100    4120    4020    4750    4020    4120    2100

25230

太庙二道门北立面图

North elevation of the Second Gate to the Ancestral Temple

0    1    3m

# 后殿及附属建筑
## The Rear Hall and Ancillary Buildings

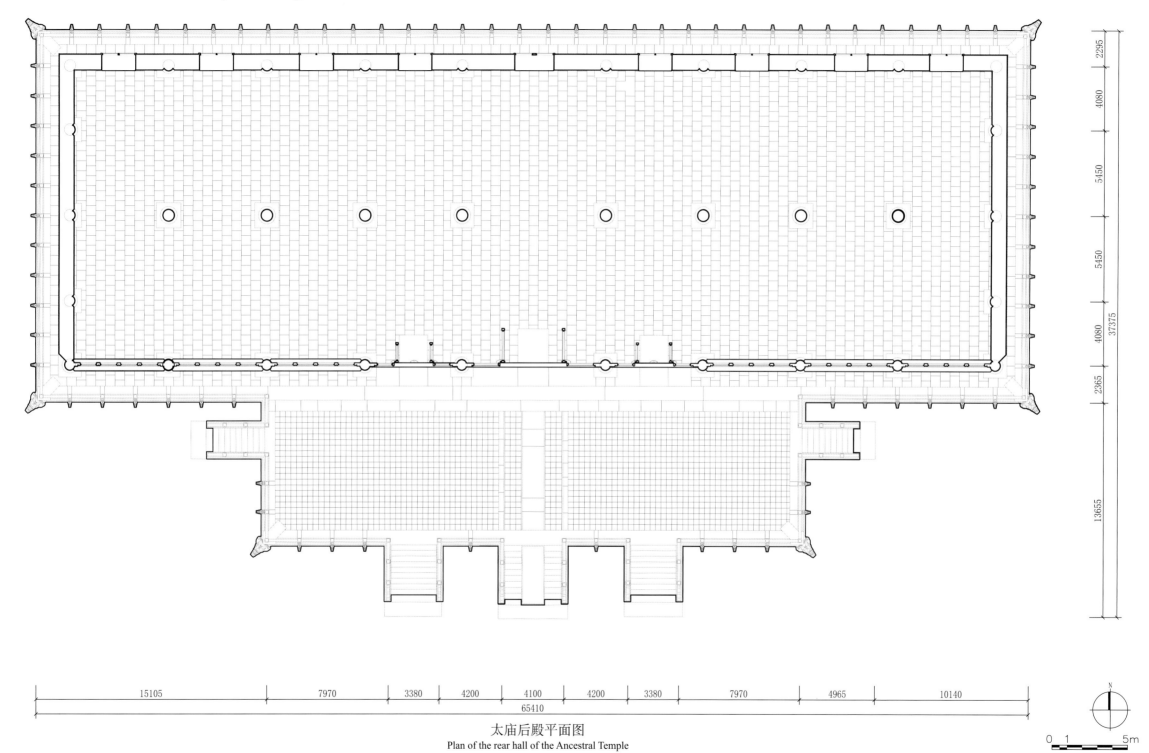

太庙后殿平面图
Plan of the rear hall of the Ancestral Temple

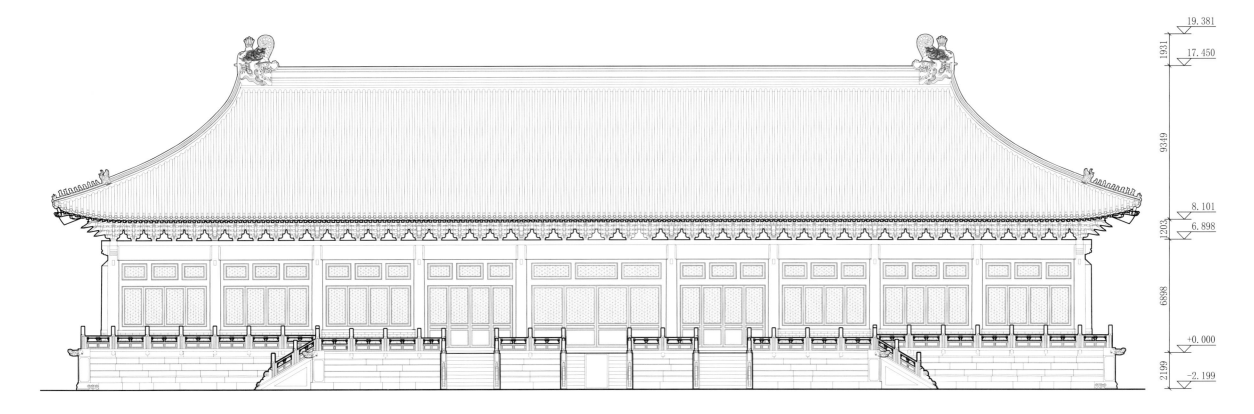

19.381

17.450

8.101

6.898

+0.000

-2.199

1931

9349

1203

6898

2199

15105    7970    3380    4200    4100    4200    3380    7970    4965    10140

65410

太庙后殿南立面图

South elevation of the rear hall of the Ancestral Temple

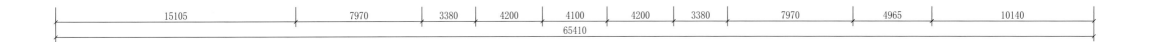

0 1     5m

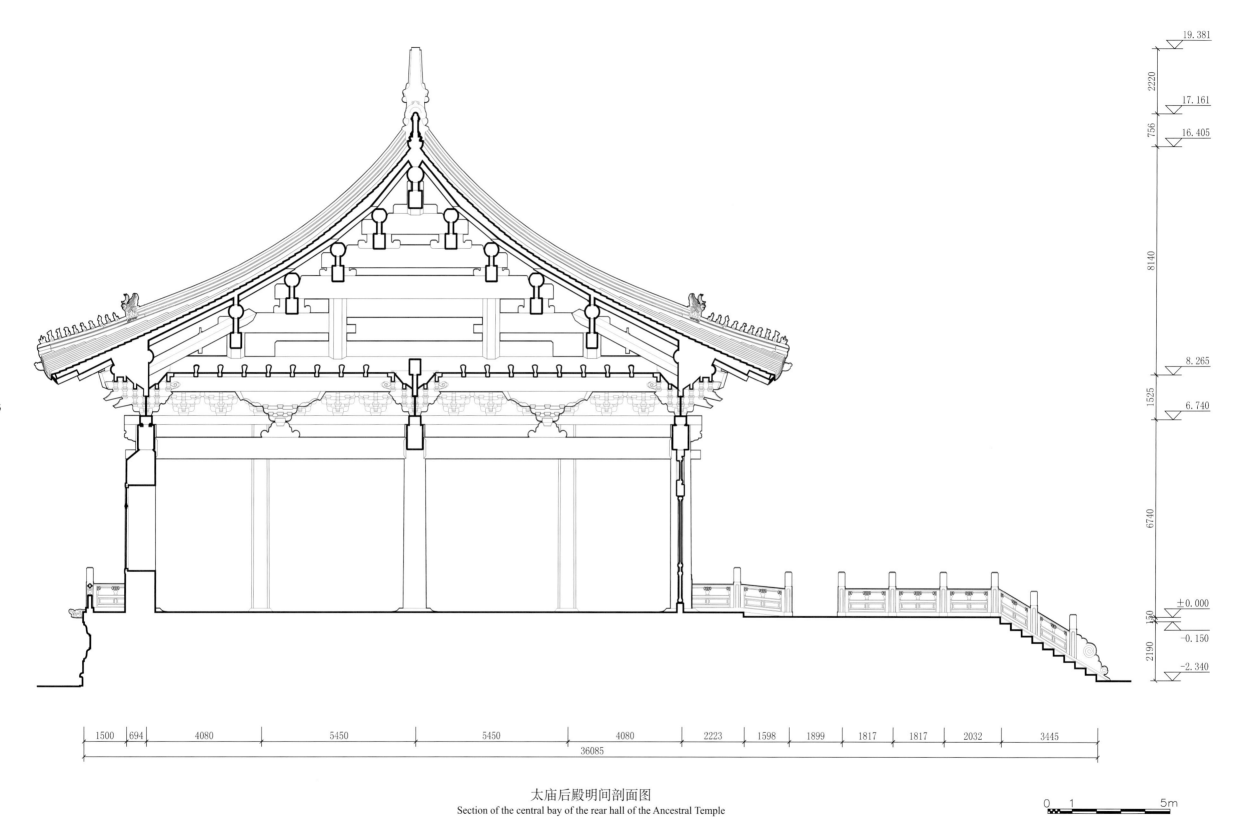

19.381

2220

17.161

756

16.405

8140

8.265

1525

6.740

6740

±0.000

130

-0.150

2190

-2.340

1500　694　4080　5450　5450　4080　2223　1598　1899　1817　1817　2032　3445

36085

太庙后殿明间剖面图
Section of the central bay of the rear hall of the Ancestral Temple

0　1　　　　5m

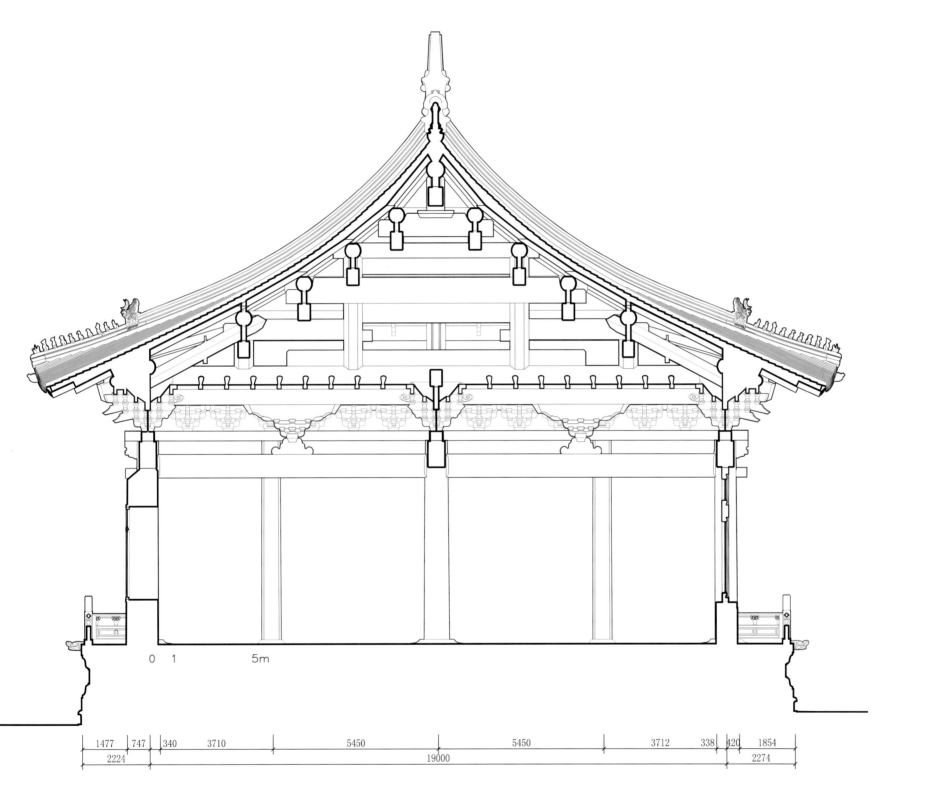

19.381

2220

17.161

756

16.405

8140

8.265

292

7.973

1543

6.430

6430

+0.000

2160

-2.160

1477 747 340 3710 5450 5450 3712 338 420 1854

2224 19000 2274

0 1 5m

太庙后殿梢间剖面图

Section of the terminal bay of the rear hall of the Ancestral Temple

0 1 5m

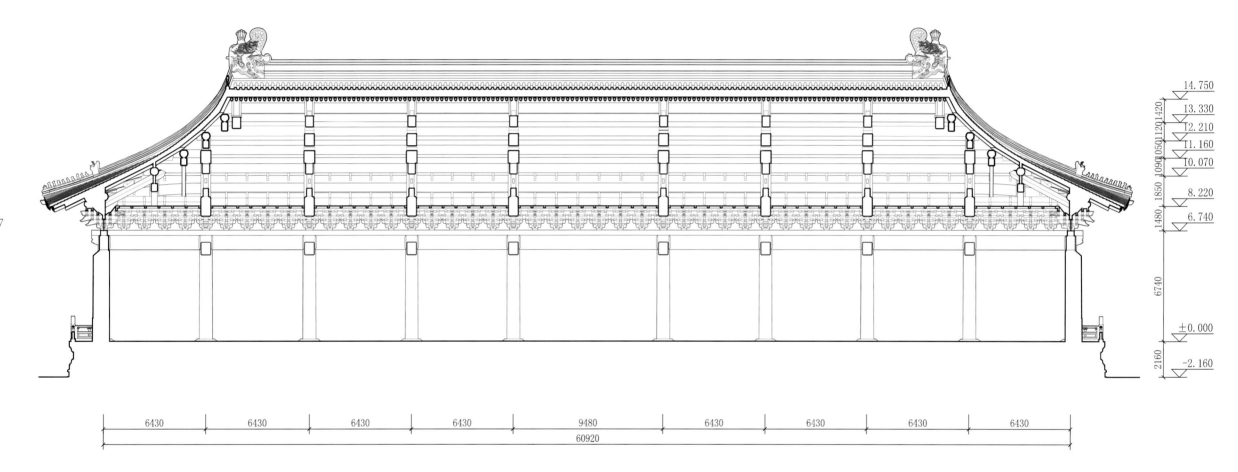

14.750
13.330
12.210
11.160
10.070
8.220
6.740
±0.000
-2.160

6430　6430　6430　6430　9480　6430　6430　6430　6430
60920

太庙后殿纵剖面图
Longitudinal section of the rear hall of the Ancestral Temple

0  1      5m

太庙后殿屋顶平面
Roof plan of the rear hall of the Ancestral Temple

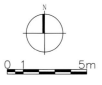

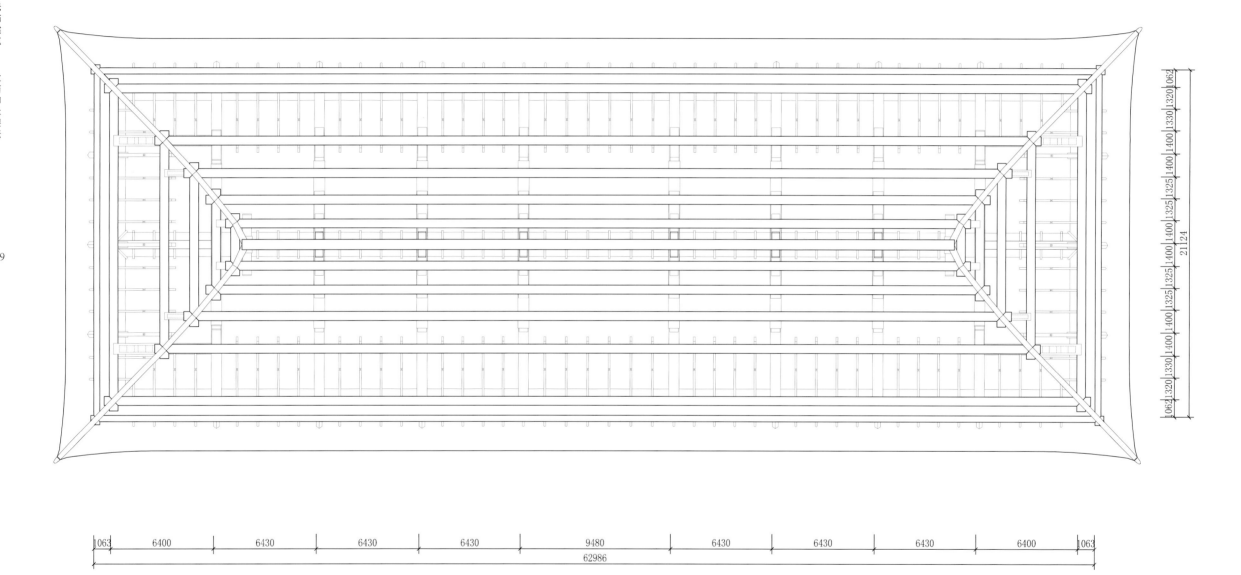

太庙后殿梁架俯视图
Top view of the beam frame of the rear hall of the Ancestral Temple

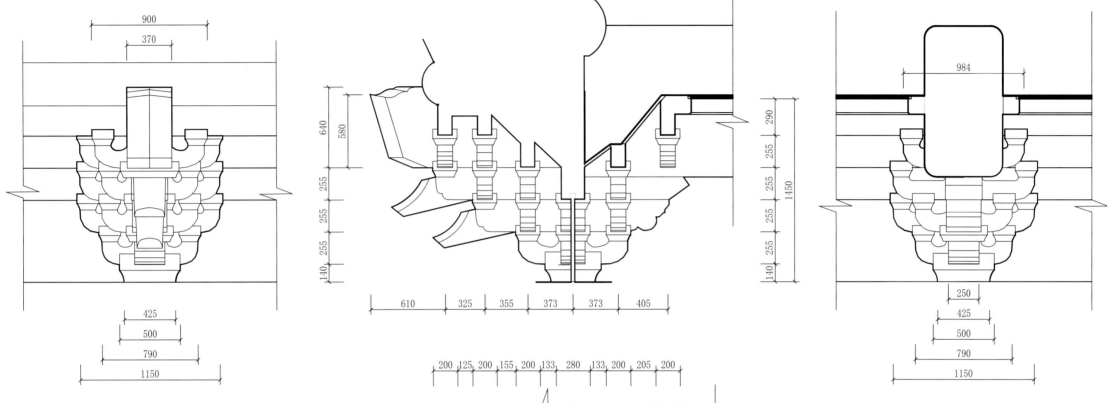

太庙后殿柱头科斗栱大样图
Detail of the connecting bracket sets on the columns of the Rear Hall of the Ancestral Temple

0    0.2    0.4m

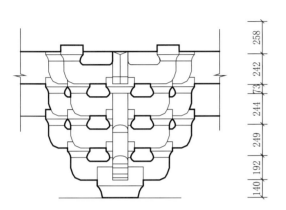

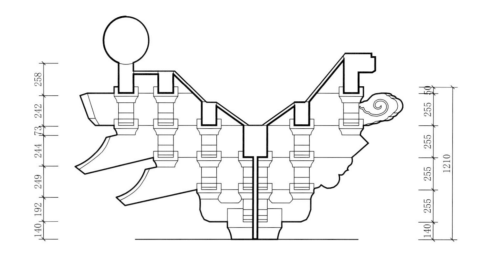

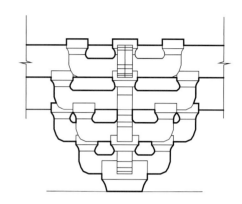

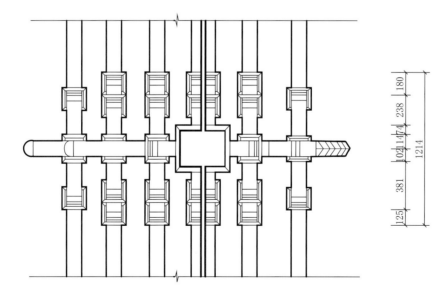

太庙后殿平身科斗栱大样图

Detail of the connecting bracket sets between the columns of the Rear Hall of the Ancestral Temple

0    0.2    0.4m

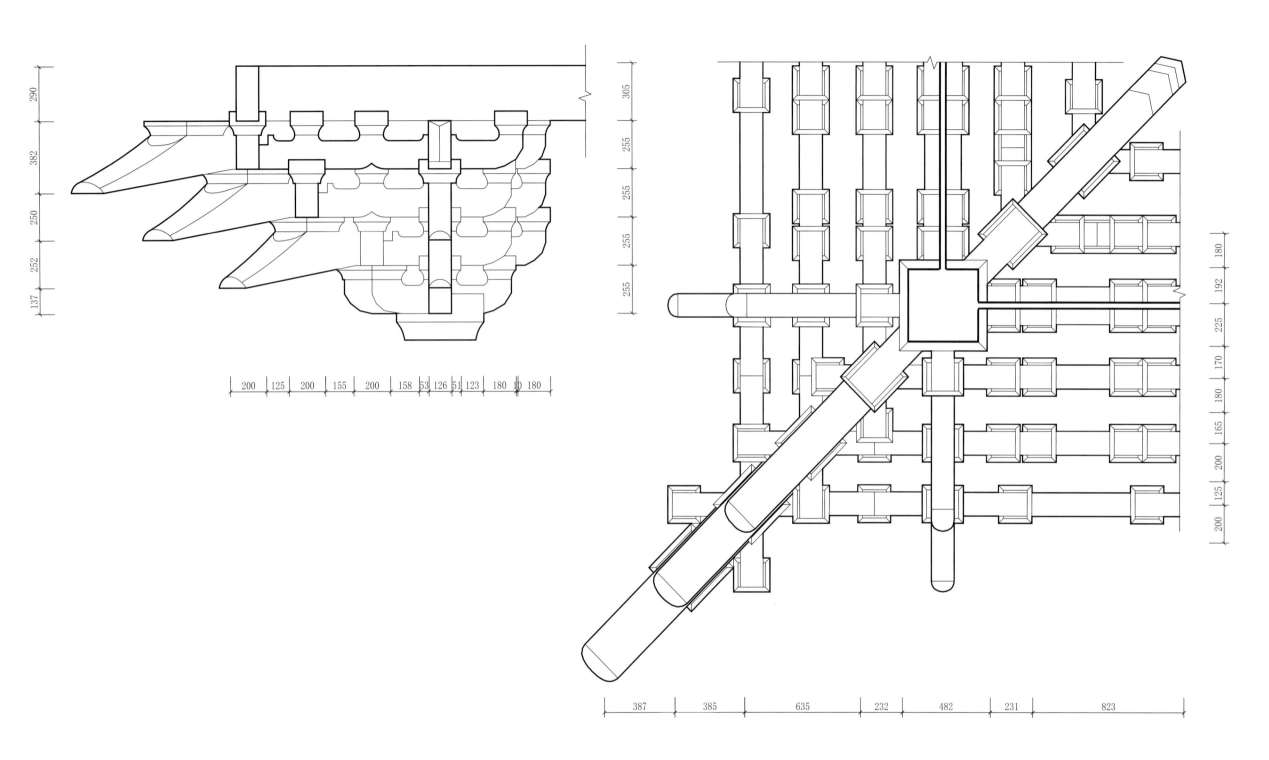

太庙后殿角科斗栱大样图

Detail of the connecting bracket sets on the corner columns of the Rear Hall of the Ancestral Temple

0    0.3    0.6m

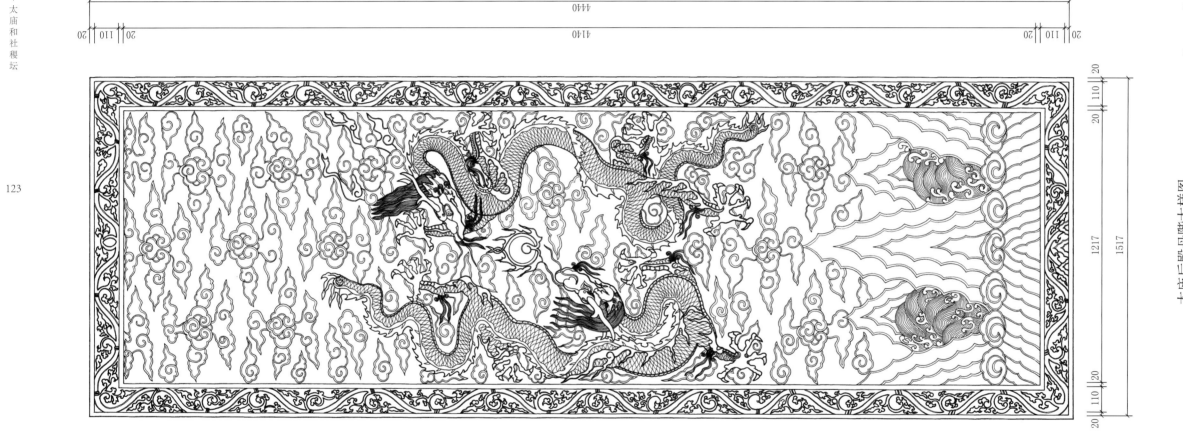

太庙后殿丹陛大样图

Front palatial steps detail drawing of the Rear Hall of the Ancestral Temple

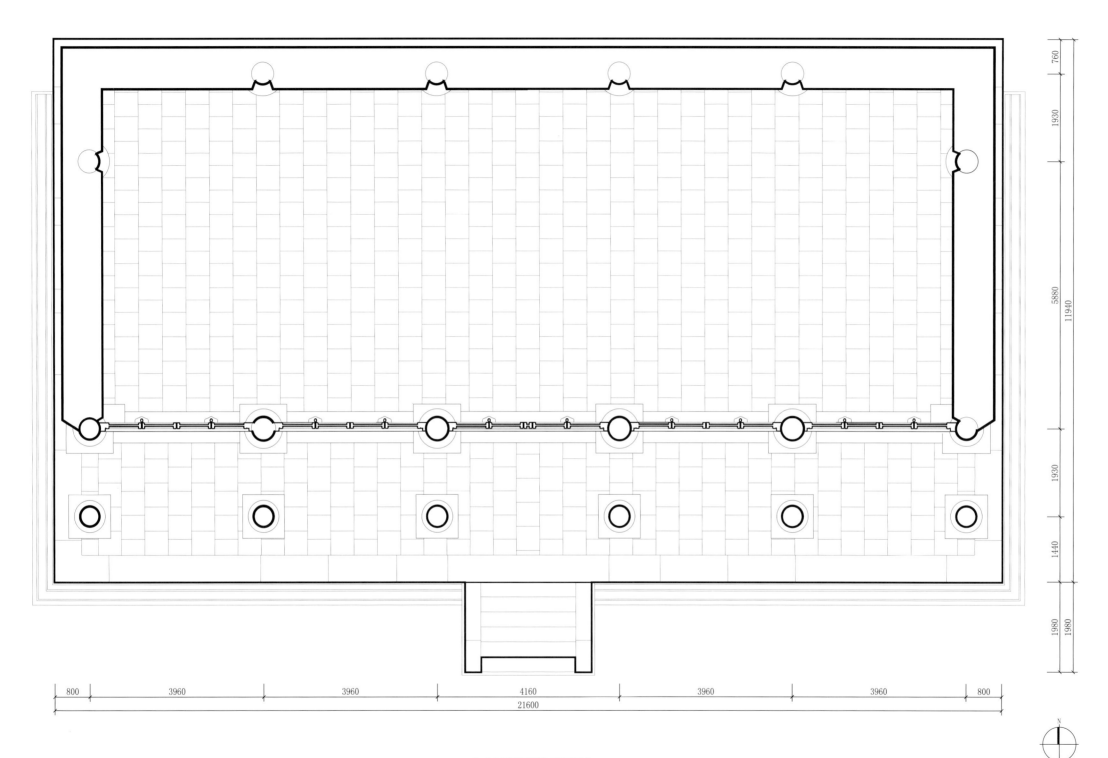

760
1930
11940 5880
1930
1440
1980 1980

800 3960 3960 4160 3960 3960 800
21600

太庙后殿配殿平面图
Plan of the side hall of the rear hall of the Ancestral Temple

0 1 2m

N

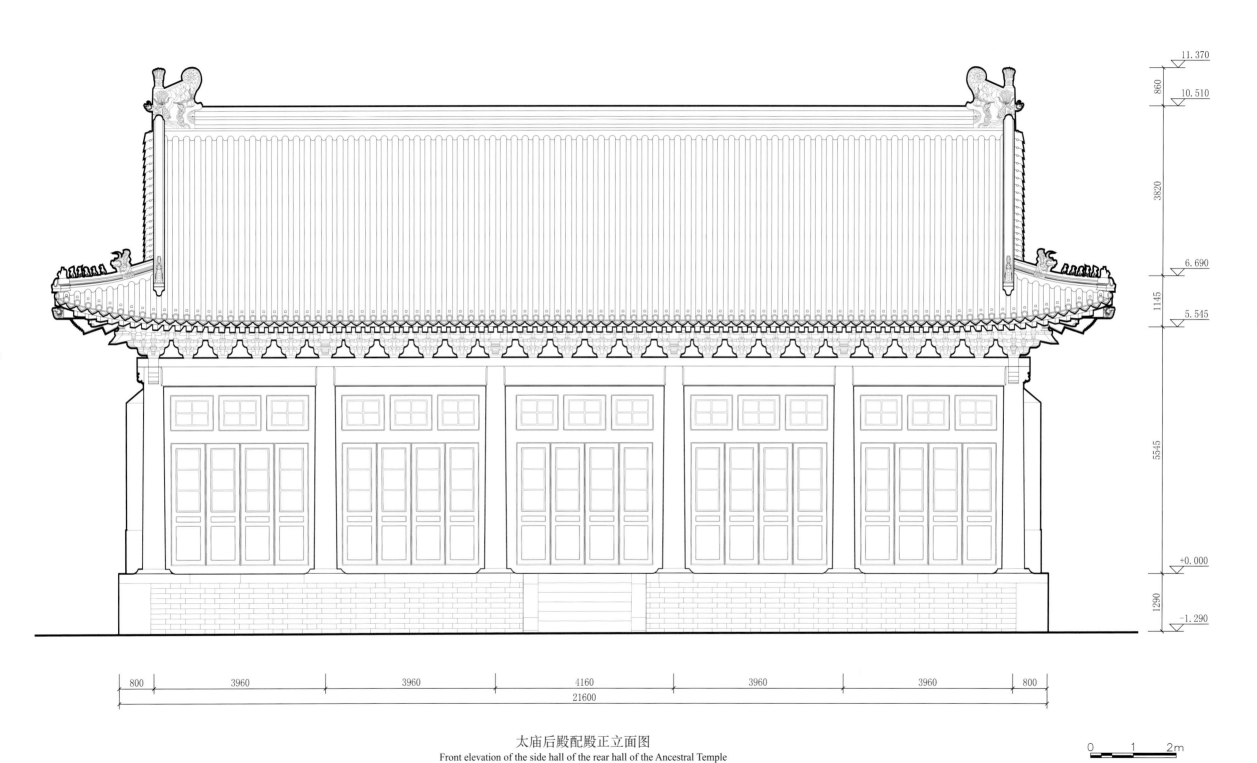

11.370

10.510

6.690

5.545

+0.000

-1.290

860

3820

1145

5545

1290

800　3960　3960　4160　3960　3960　800

21600

太庙后殿配殿正立面图

Front elevation of the side hall of the rear hall of the Ancestral Temple

0　1　2m

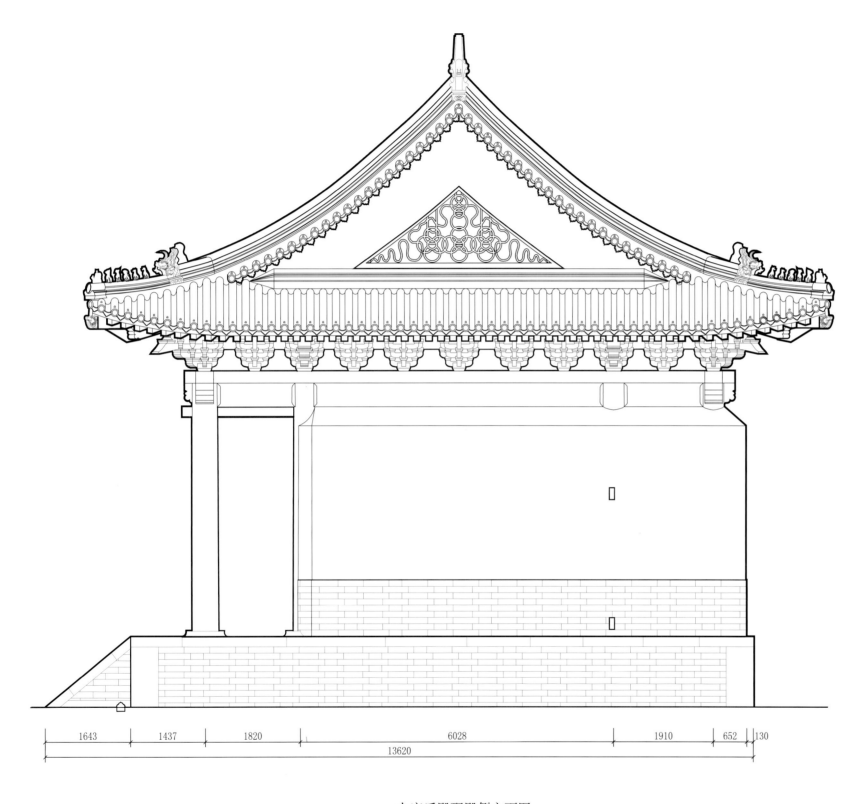

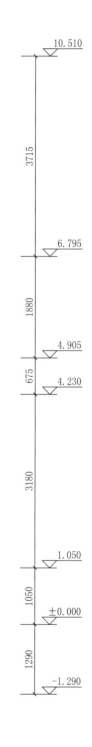

10.510

3715

6.795

1880

4.905

675

4.230

3180

1.050

1050

±0.000

1290

-1.290

| 1643 | 1437 | 1820 | 6028 | 1910 | 652 | 130 |

13620

太庙后殿配殿侧立面图
Side elevation of the side hall of the rear hall of the Ancestral Temple

0  0.5  1        2m

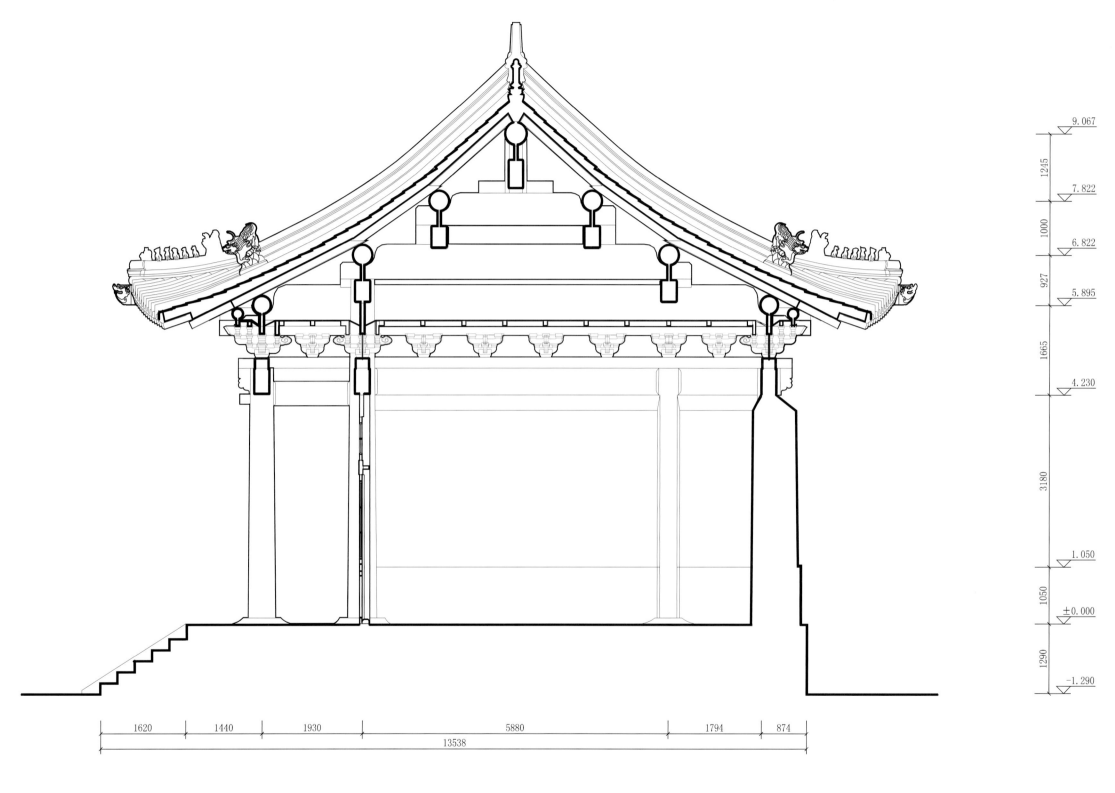

9.067

1245

7.822

1000

6.822

927

5.895

1665

4.230

3180

1.050

1050

±0.000

1290

-1.290

1620　1440　1930　5880　1794　874

13538

太庙后殿配殿明间剖面图
Section of the central bay of the side hall of the rear hall of the Ancestral Temple

0　　1　　2m

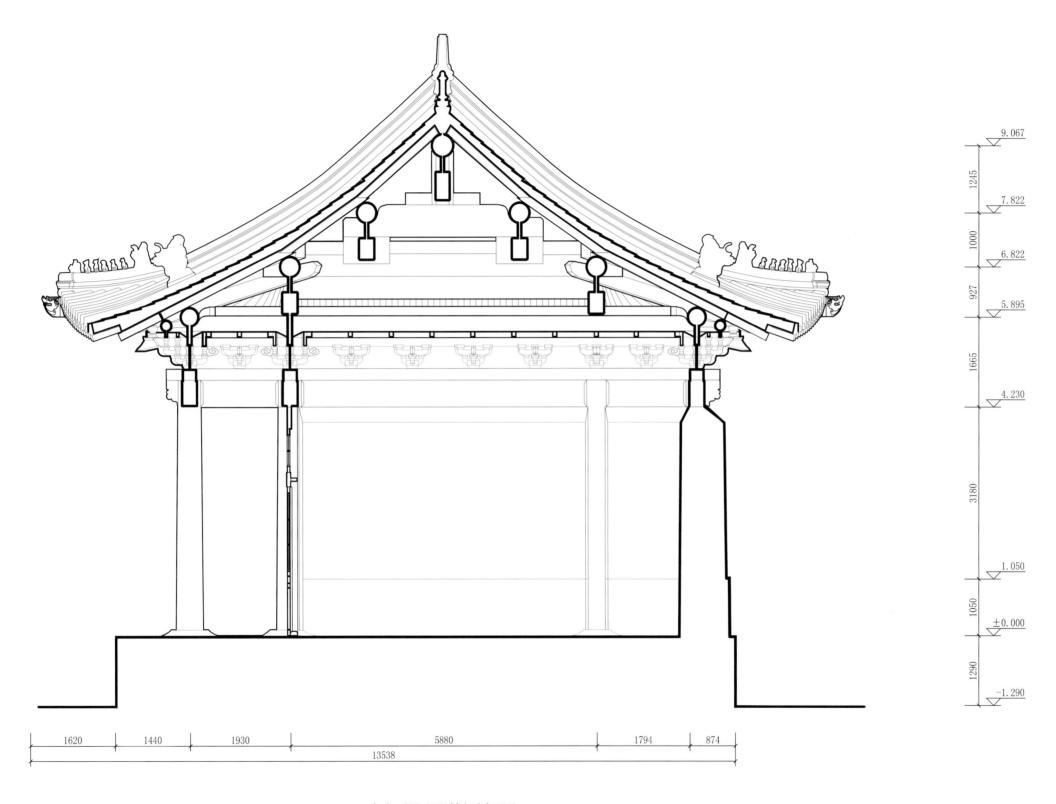

9.067

1245

7.822

1000

6.822

927

5.895

1665

4.230

3180

1.050

1050

±0.000

1290

-1.290

1620　1440　1930　5880　1794　874

13538

太庙后殿配殿梢间剖面图
Section of the terminal bay of the side hall of the rear hall of the Ancestral Temple

0　1　2m

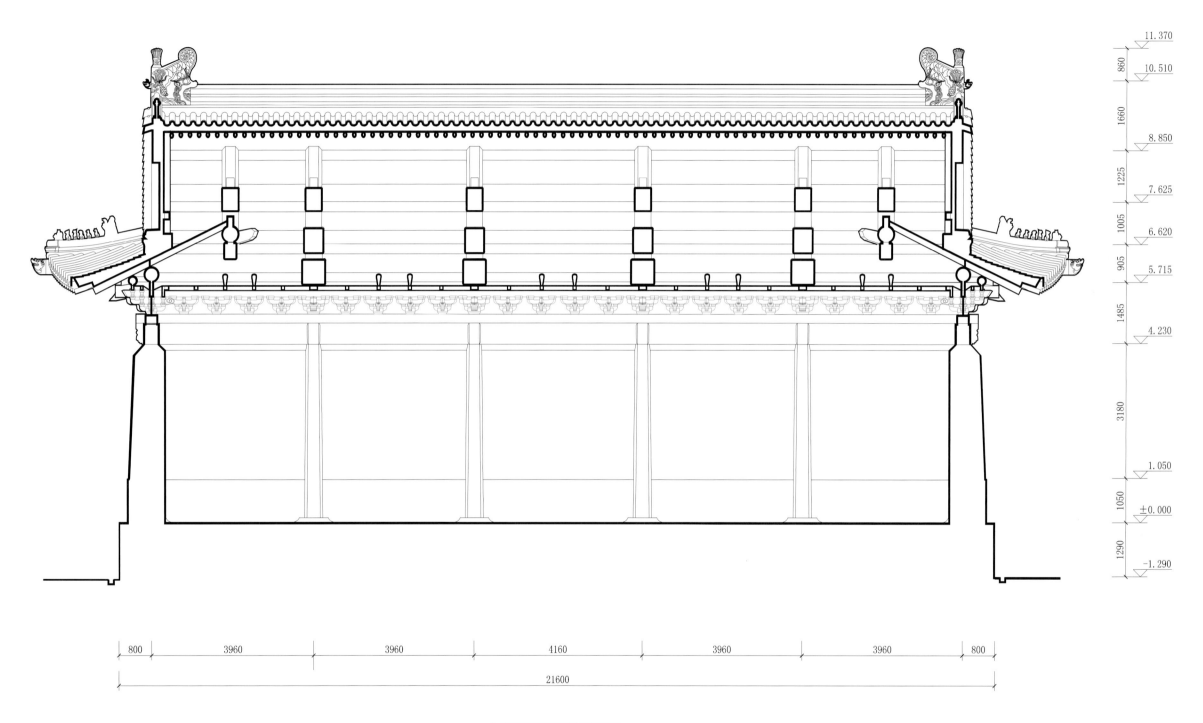

太庙后殿配殿纵剖面图
Longitudinal section of the side hall of the rear hall of the Ancestral Temple

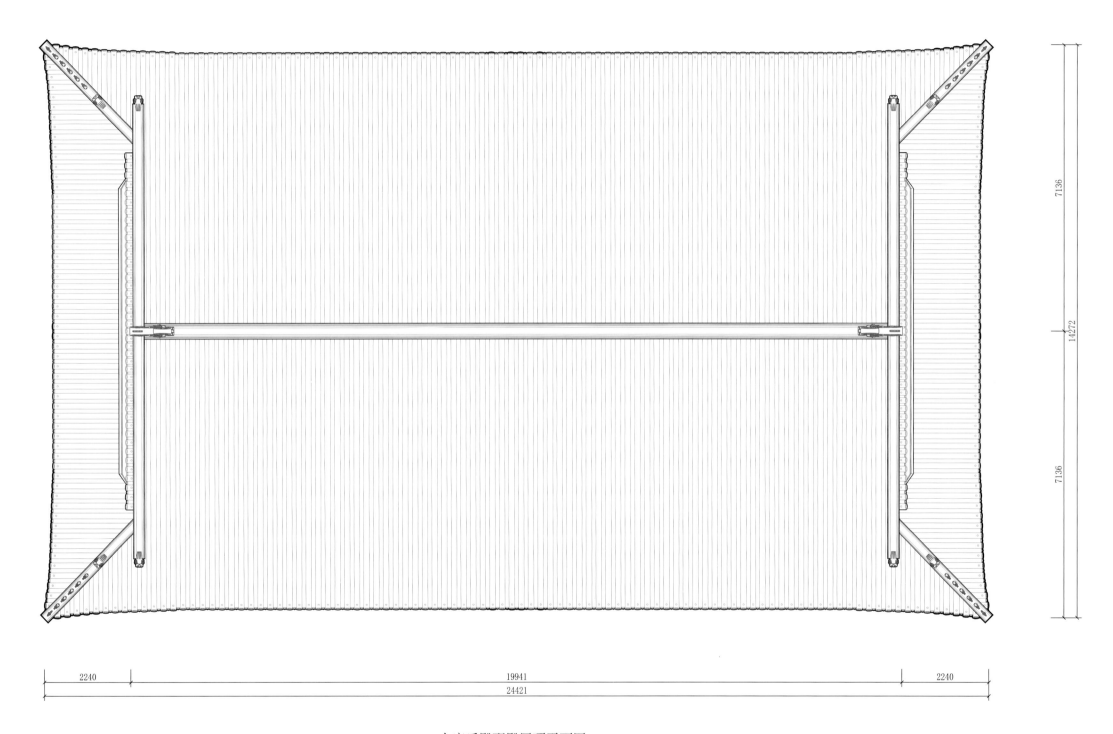

太庙后殿配殿屋顶平面图
Roof plan of the side hall of the rear hall of the Ancestral Temple

0 0.5 1 2m

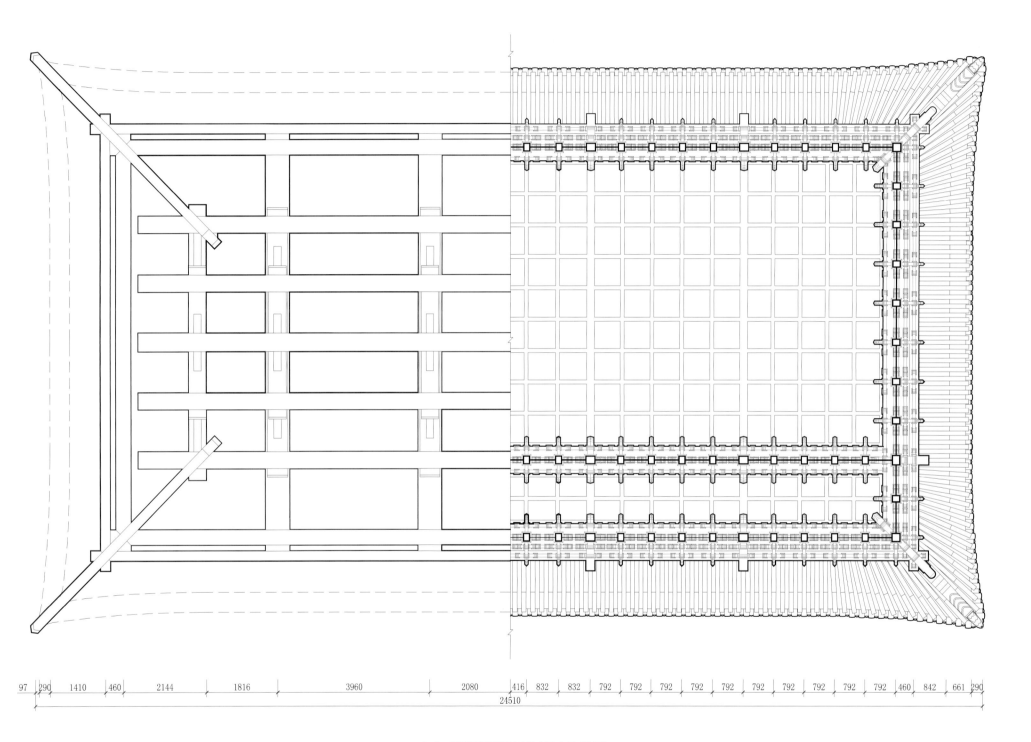

太庙后殿配殿梁架俯视及仰视图
Top view of the beam frame of the side hall of the rear hall of the Ancestral Temple

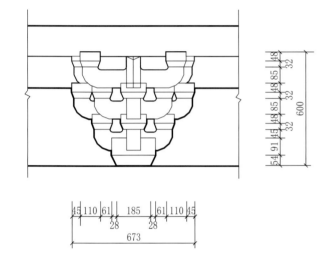

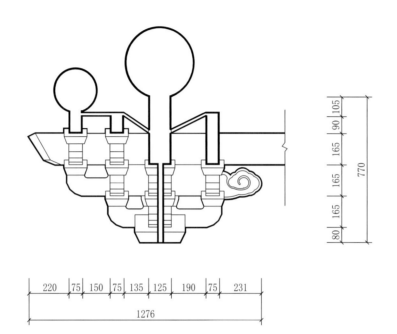

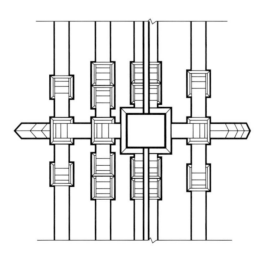

太庙后殿配殿平身科斗栱大样图

Detail of the connecting bracket sets between the columns of the Side Hall of the Rear Hall of the Ancestral Temple

0    0.2    0.4m

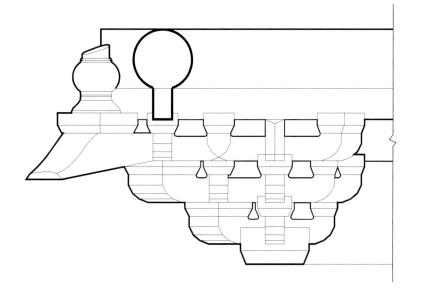
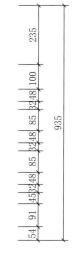
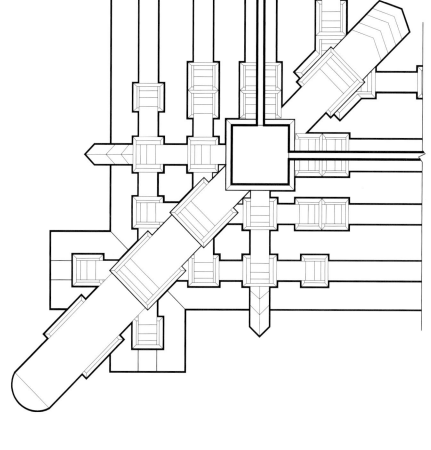

太庙后殿配殿角科斗栱大样图

Detail of the connecting bracket sets on the corner columns of the Side Hall of the Rear Hall of the Ancestral Temple

0　0.15　0.3m

中国古建筑测绘大系·坛庙建筑 —— 太庙和社稷坛

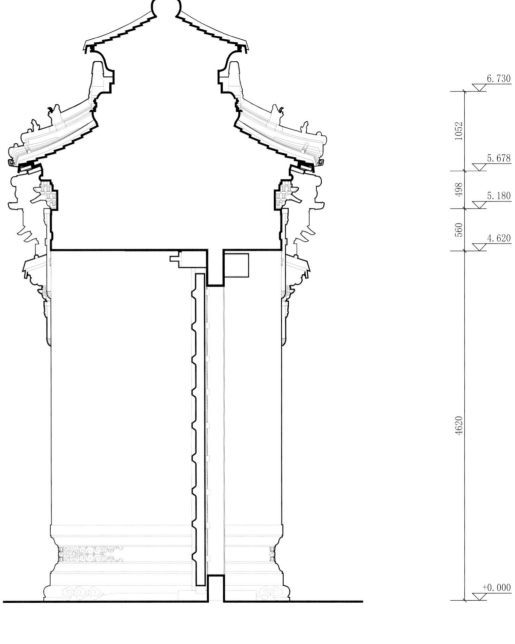

8.220

823

7.397

2108

5.296

1341

3.955

2957

0.998

998

+0.000

6.730

1052

5.678

498

5.180

560

4.620

4620

+0.000

1100　1100
2200

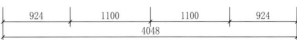

924　1100　1100　924
4048

太庙三道门侧立面图
Side elevation of the Third Gate of the Ancestral Temple

太庙三道门中门剖面图
Middle entrance section of the Third Gate of the Ancestral Temple

0　1　　　　5m

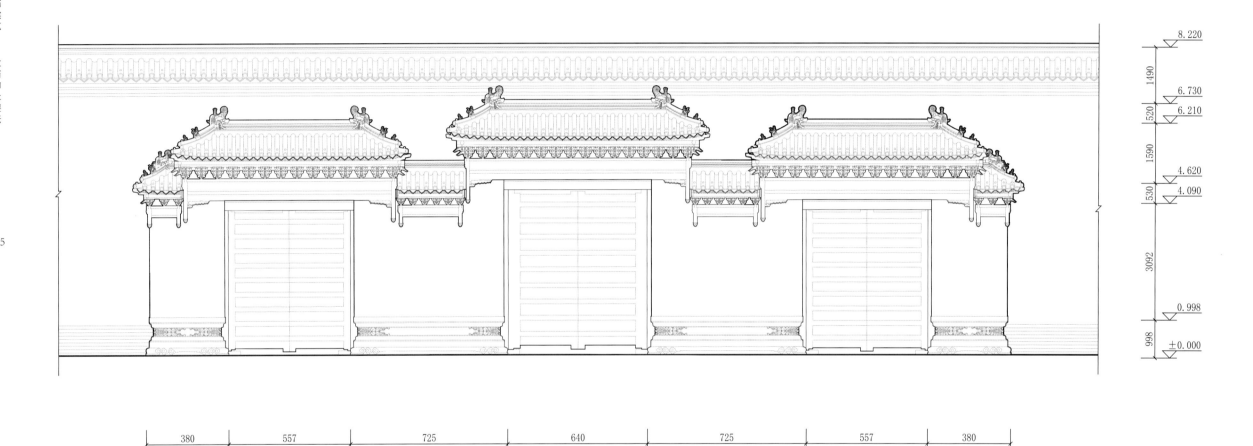

太庙三道门北立面图

North elevation of the Third Gate of the Ancestral Temple

0 1    5m

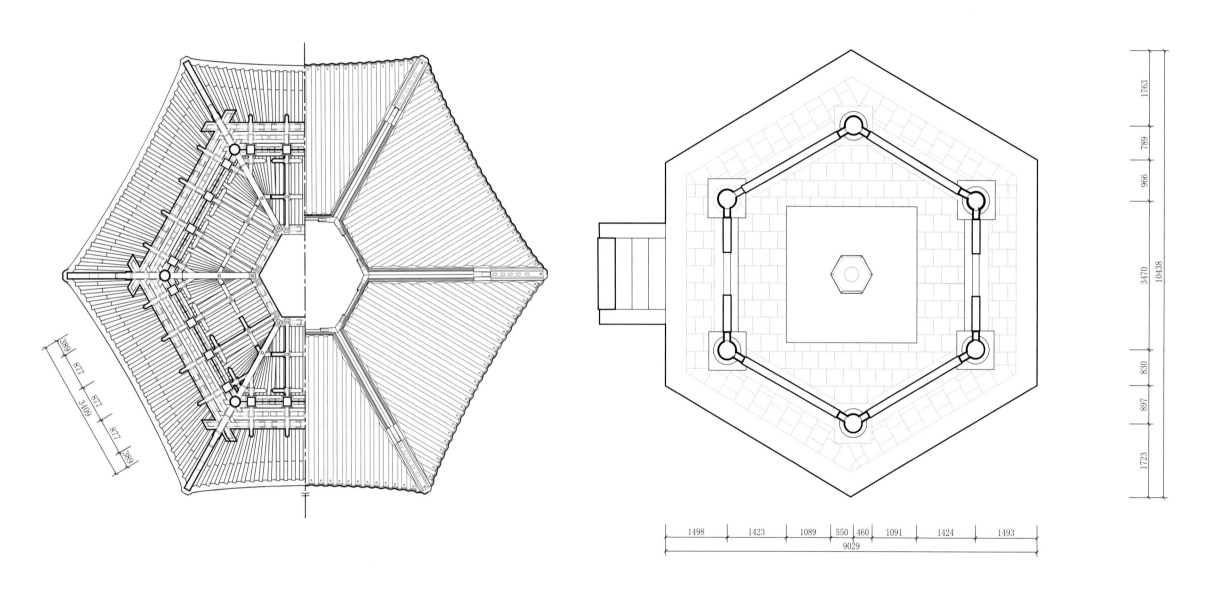

太庙井亭屋顶平面图及梁架仰视图
Roof plan and bottom view of the beam frame of the Well Pavilion of the Ancestral Temple

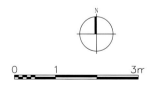

太庙井亭平面图
Plan of the Well Pavilion of the Ancestral Temple

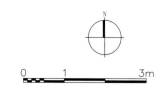

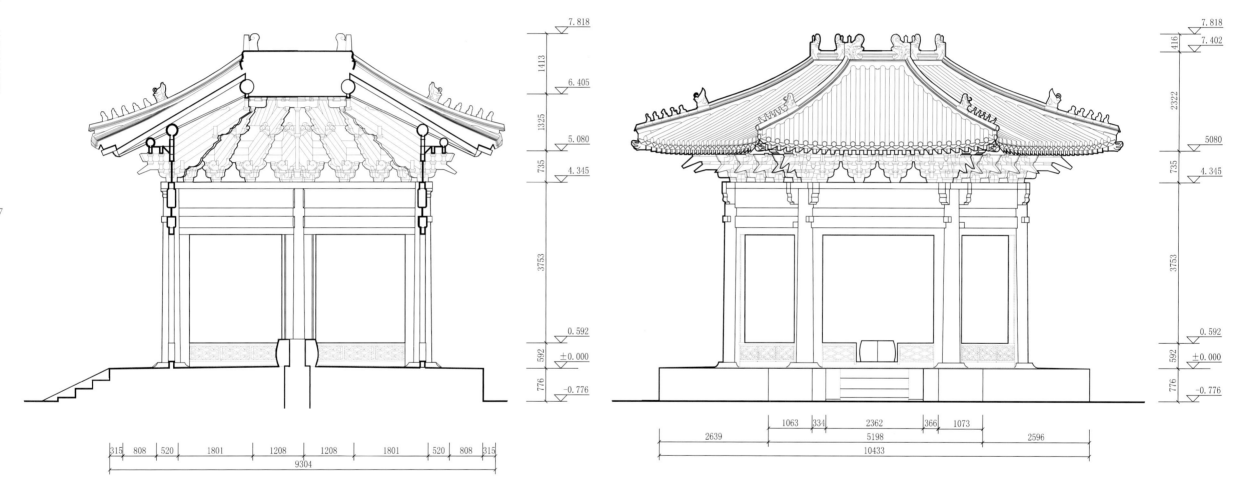

太庙井亭剖面图
Section of the Well Pavilion of the Ancestral Temple

太庙井亭西立面图
West elevation of the Well Pavilion of the Ancestral Temple

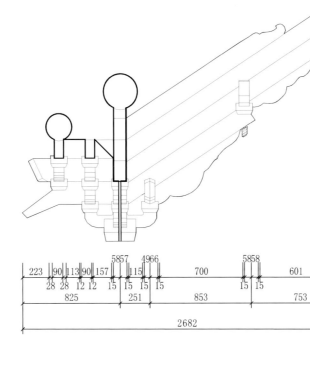

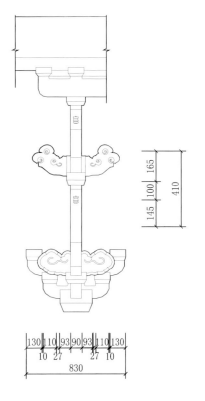

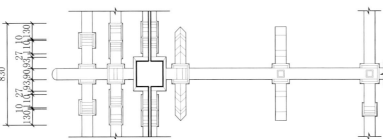

太庙井亭平身科斗栱大样

Detail of the connecting bracket sets between the columns of the Well Pavilion of the Ancestral Temple

0    0.3    0.6m

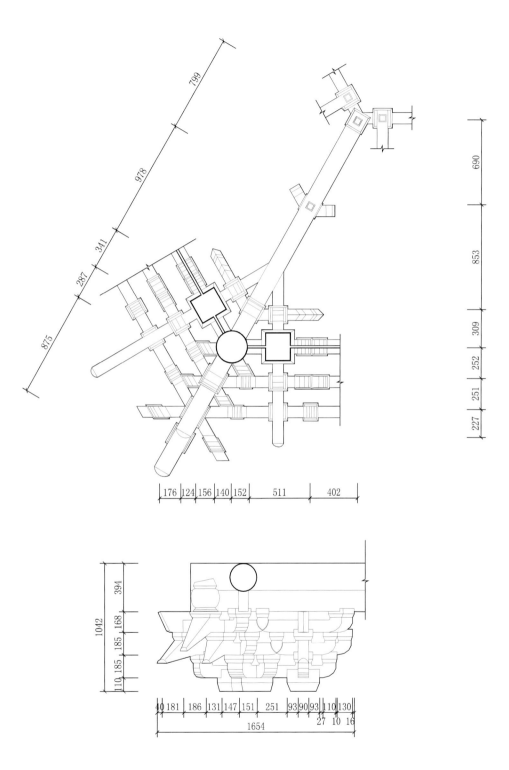

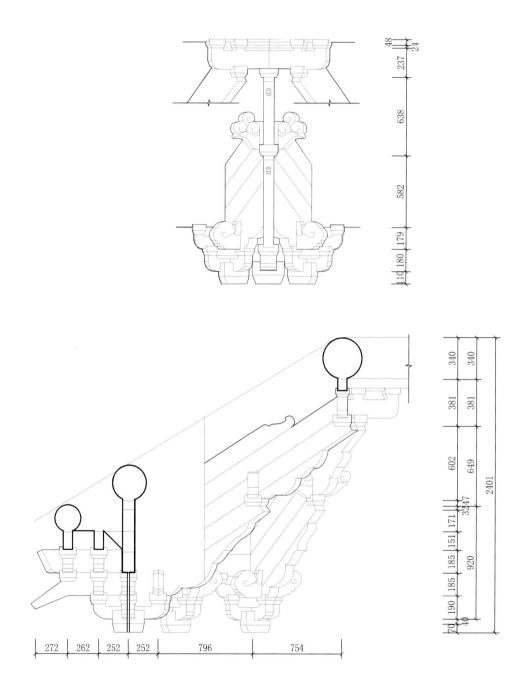

太庙井亭柱头科斗栱大样

Detail of the connecting bracket sets on the columns of the Well Pavilion of the Ancestral Temple

0    0.3    0.6m

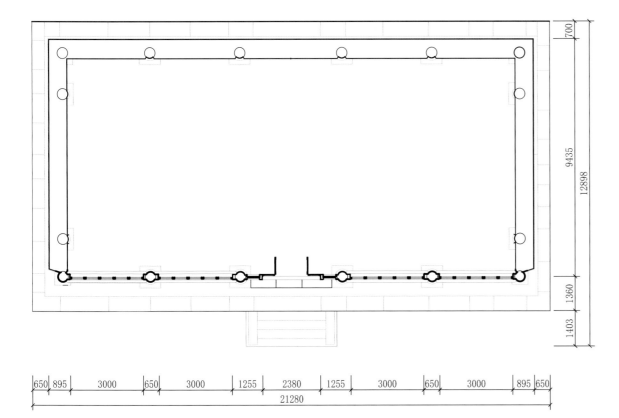

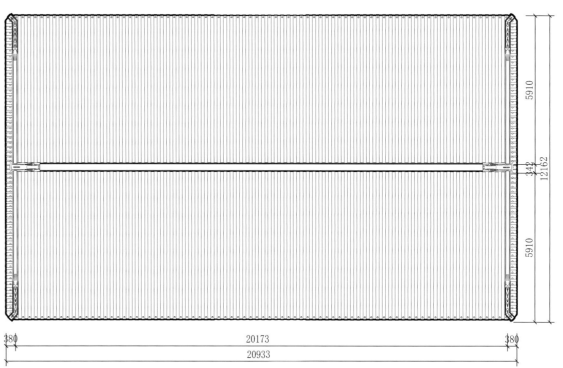

太庙神厨平面图
Plan of the Divine Kitchen of the Ancestral Temple

太庙神厨顶平面图
Roof plan of the Divine Kitchen of the Ancestral Temple

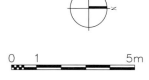

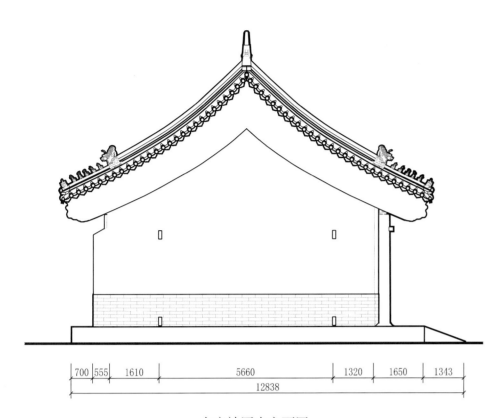

太庙神厨南立面图
South elevation of the Divine Kitchen of the Ancestral Temple

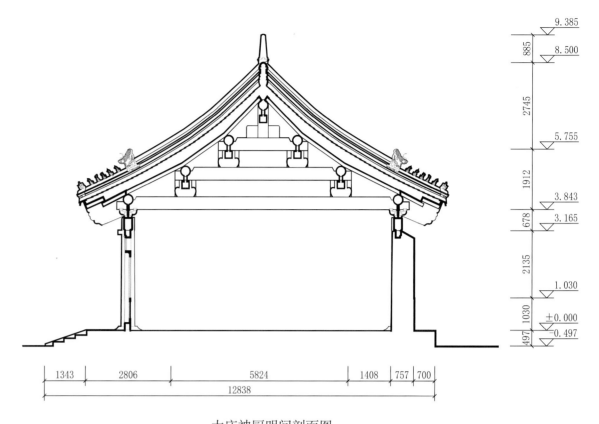

太庙神厨明间剖面图
Section the Central bay of the Divine Kitchen of the Ancestral Temple

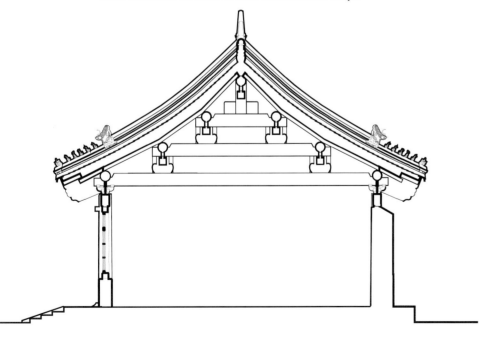

太庙神厨次间剖面图
Section the Side bay of the Divine Kitchen of the Ancestral Temple

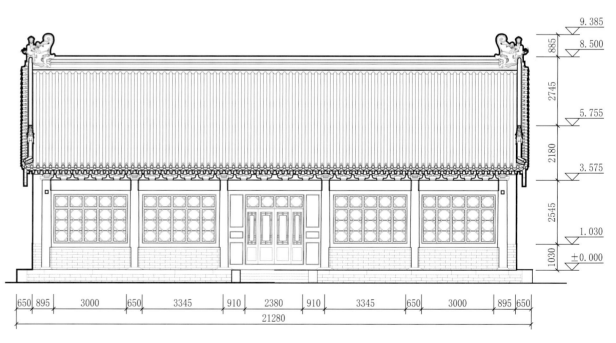

太庙神厨东立面图
East elevation of the Divine Kitchen of the Ancestral Temple

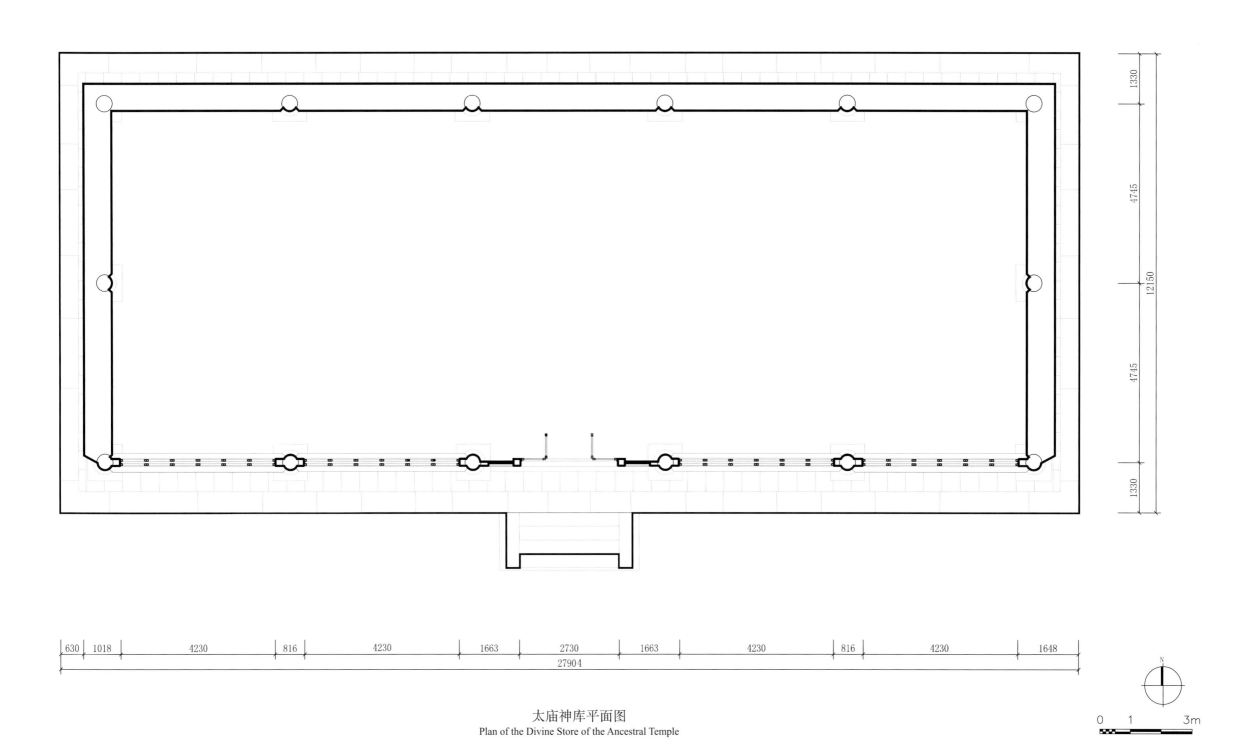

太庙神库平面图
Plan of the Divine Store of the Ancestral Temple

0  1  3m

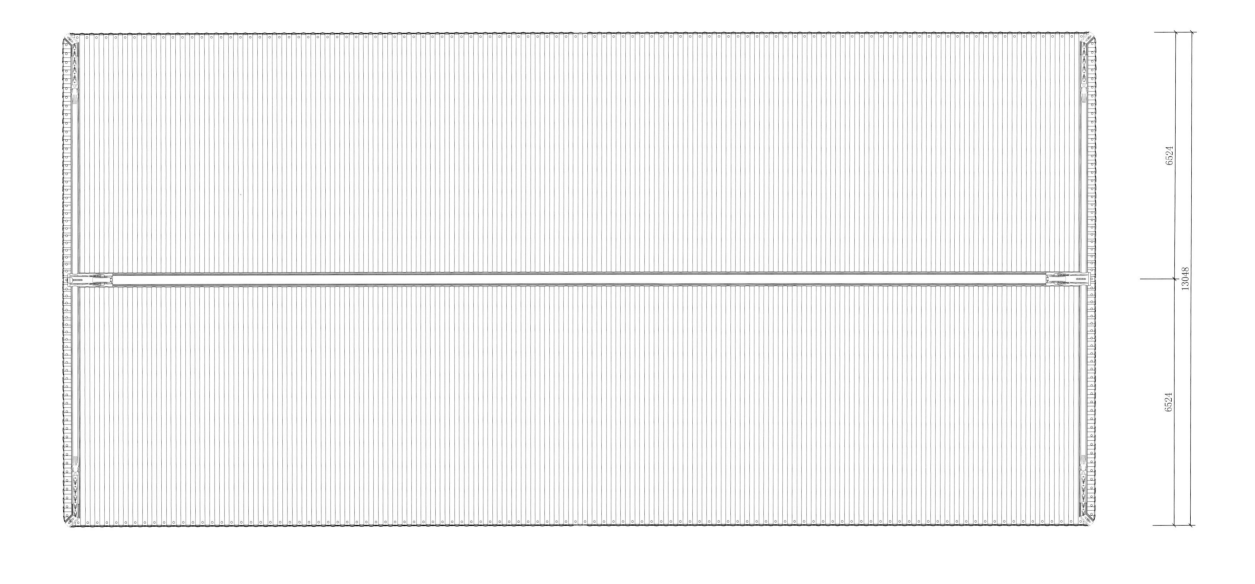

6524

13048

6521

28258

太庙神库屋顶平面图
Roof plan of the Divine Store of the Ancestral Temple

0 1 3m

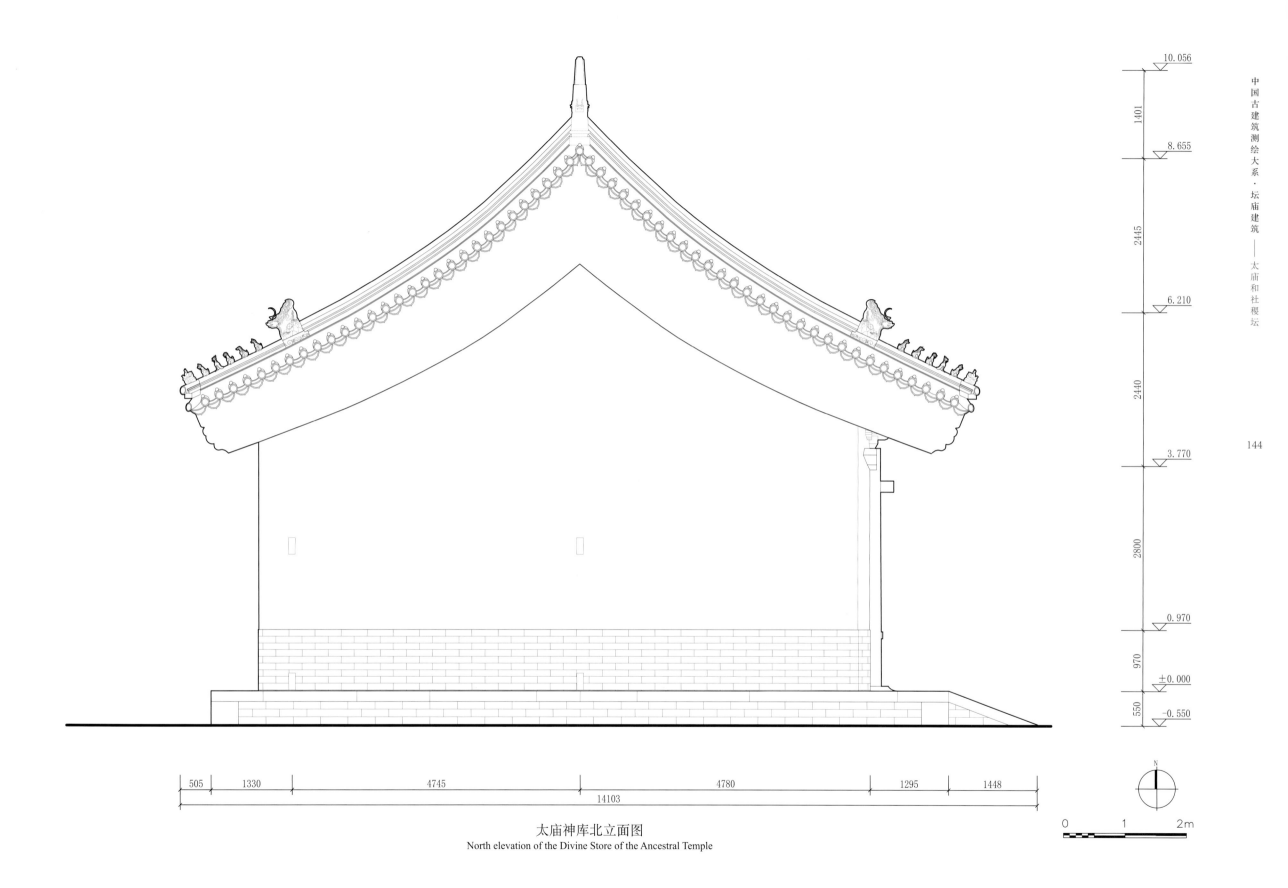

太庙神库北立面图
North elevation of the Divine Store of the Ancestral Temple

10.056

1401

8.655

2445

6.210

2440

3.770

2800

0.970

970

±0.000

550

−0.550

505　1330　4745　4780　1295　1448

14103

N

0　1　2m

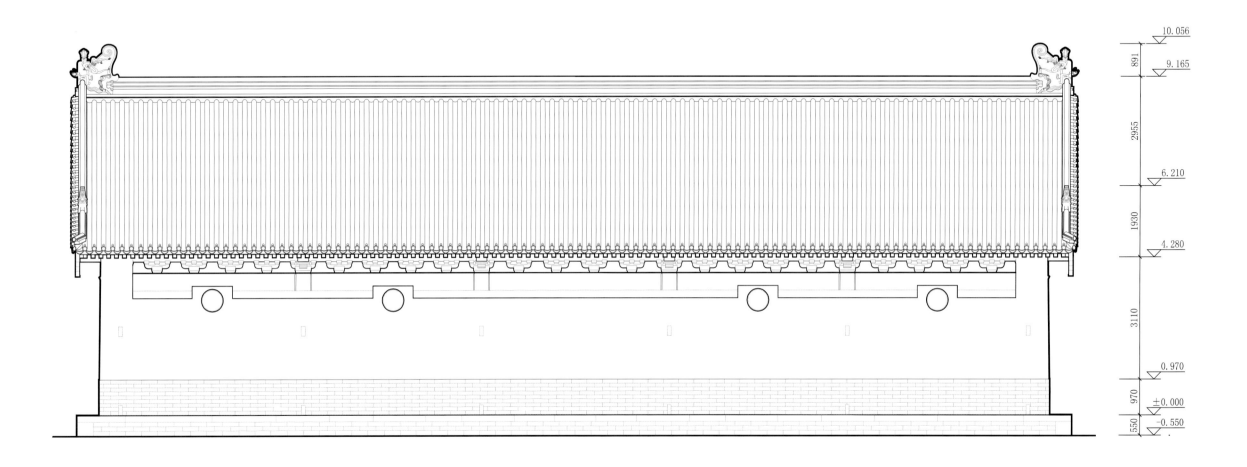

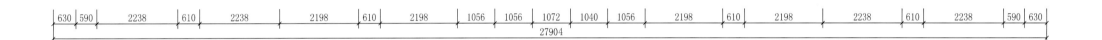

太庙神库东立面图
East elevation of the Divine Store of the Ancestral Temple

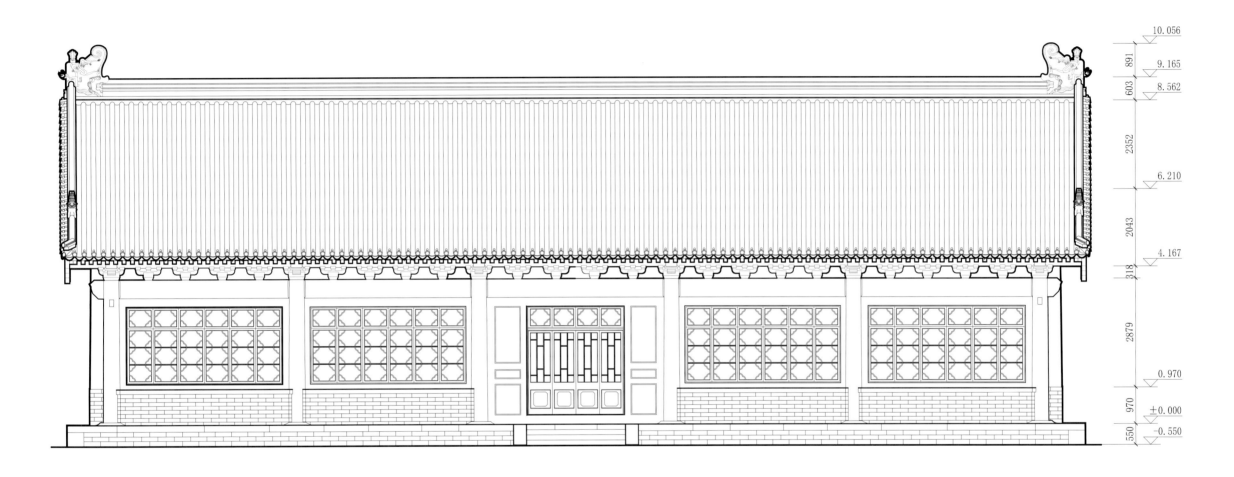

10.056
9.165
8.562
6.210
4.167
0.970
±0.000
-0.550

891
603
2352
2043
318
2879
970
550

630 590 428  4230  428 388  4230  388 1325  665 650 650 665  1325 388  4230  388 428  4230  428 590 630
27904

太庙神库西立面图
West elevation of the Divine Store of the Ancestral Temple

0 1 3m

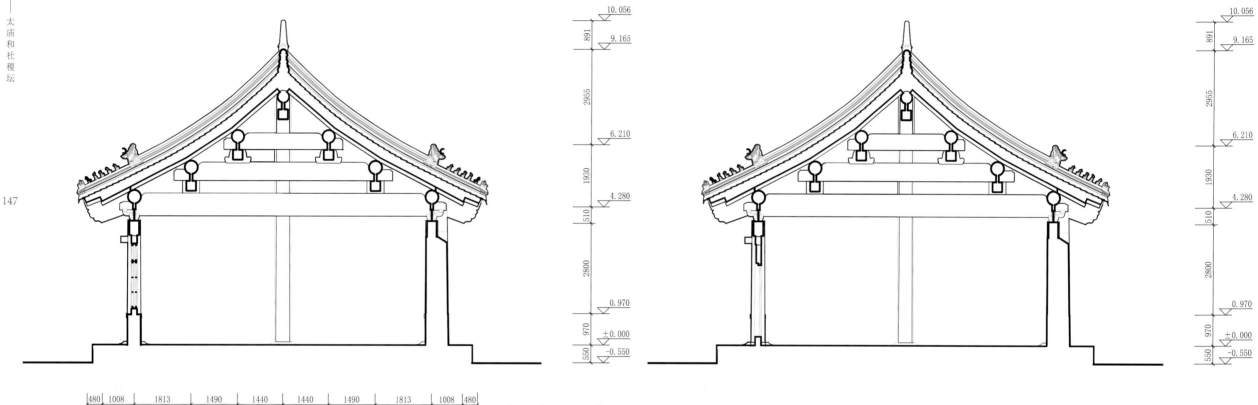

太庙神库次间剖面图
Sections of the side bay of the Divine Store of the Ancestral Temple

太庙神库明间剖面图
Sections of the central bay of the Divine Store of the Ancestral Temple

社稷坛

The Altar of
Land and Grain

中国古建筑测绘大系·坛庙建筑——太庙和社稷坛

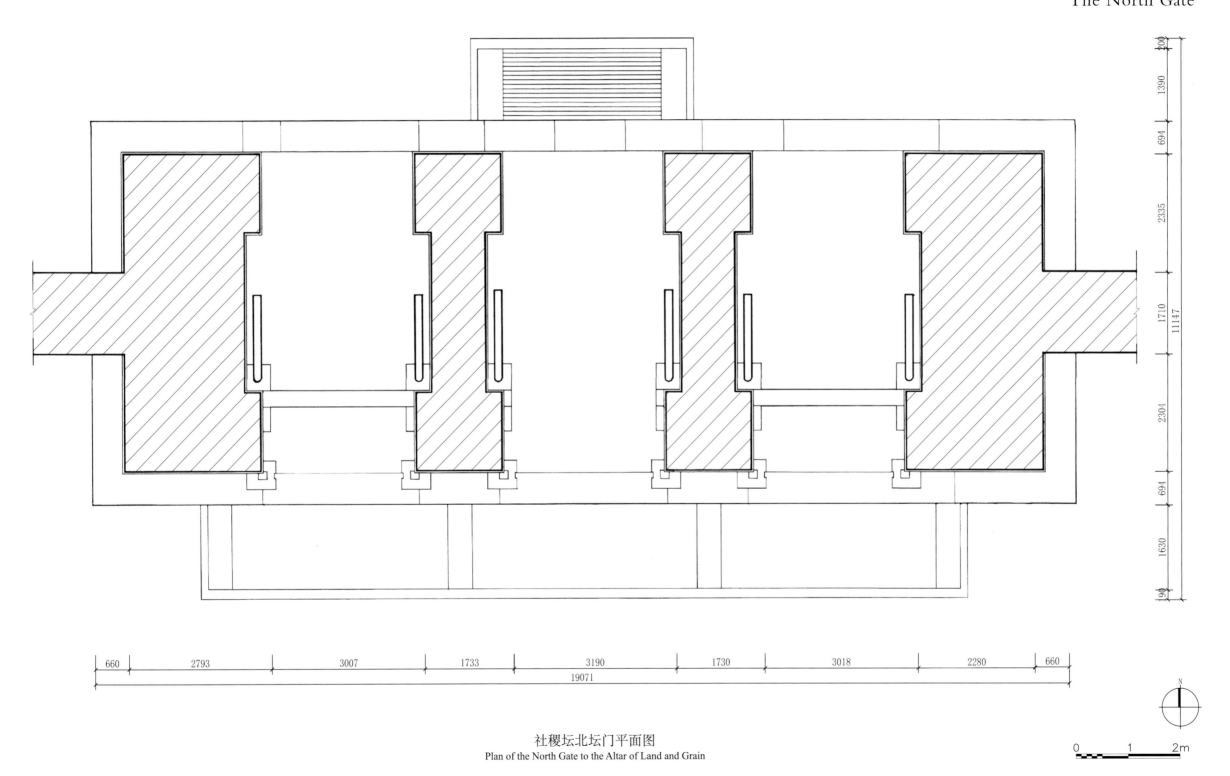

150

社稷坛北坛门平面图
Plan of the North Gate to the Altar of Land and Grain

0 1 2m

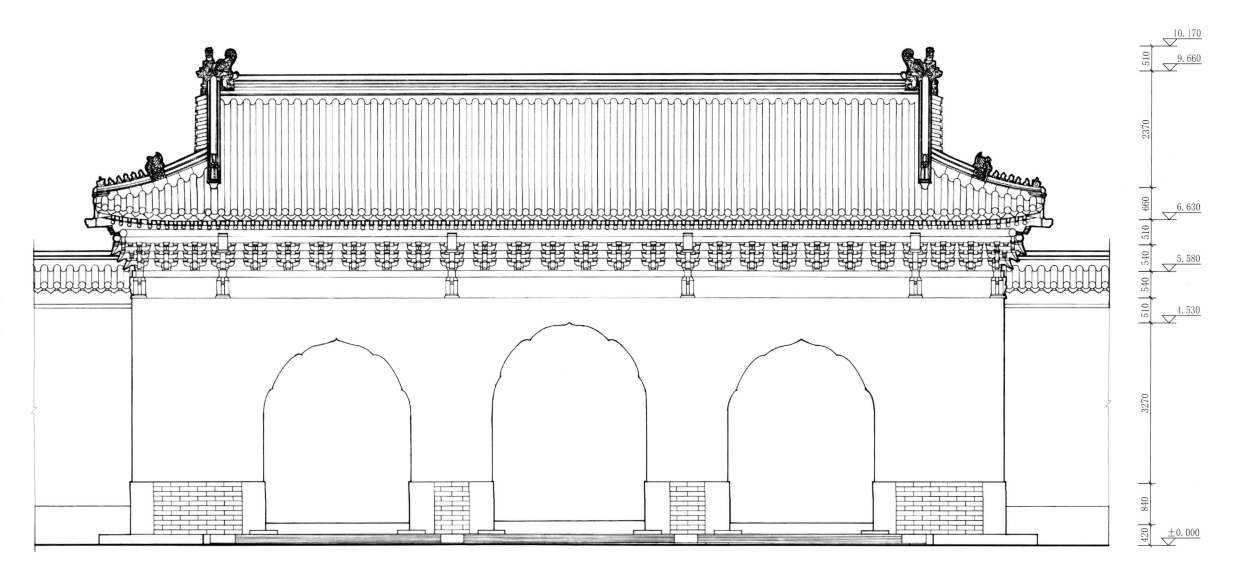

社稷坛北坛门立面图
Elevation of the North Gate to the Altar of Land and Grain

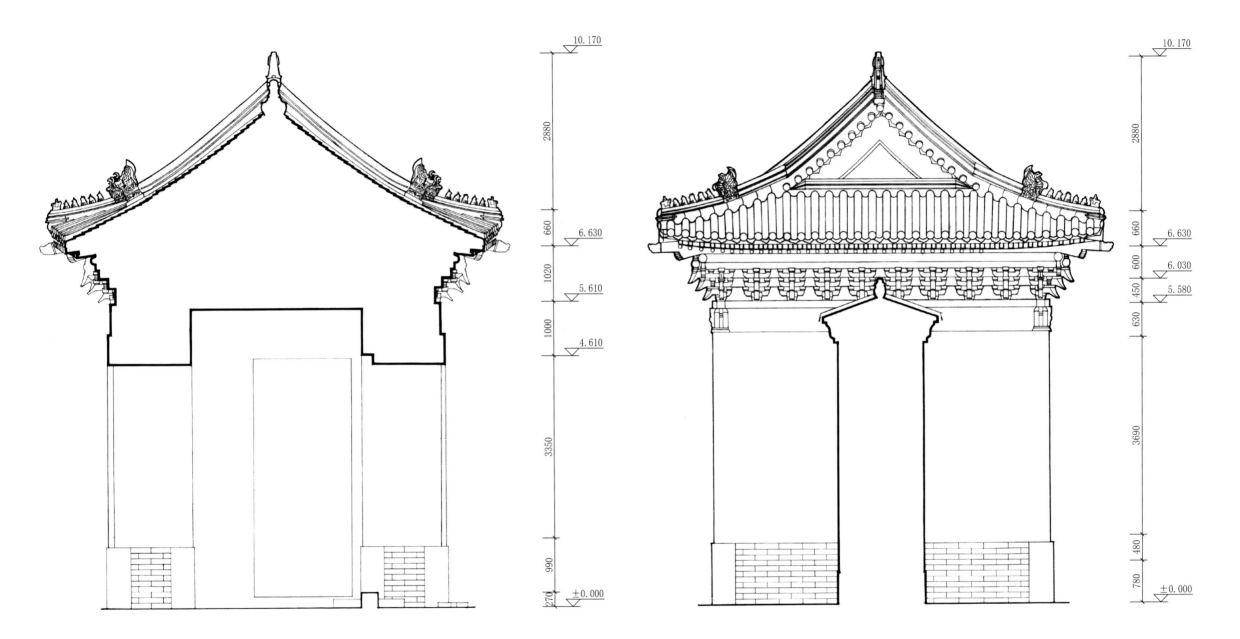

社稷坛北坛门纵剖面图
Longitudinal section of the North Gate to the Altar of Land and Grain

社稷坛北坛门侧立面图
Side elevation of the North Gate to the Altar of Land and Grain

0　　　1　　　2m

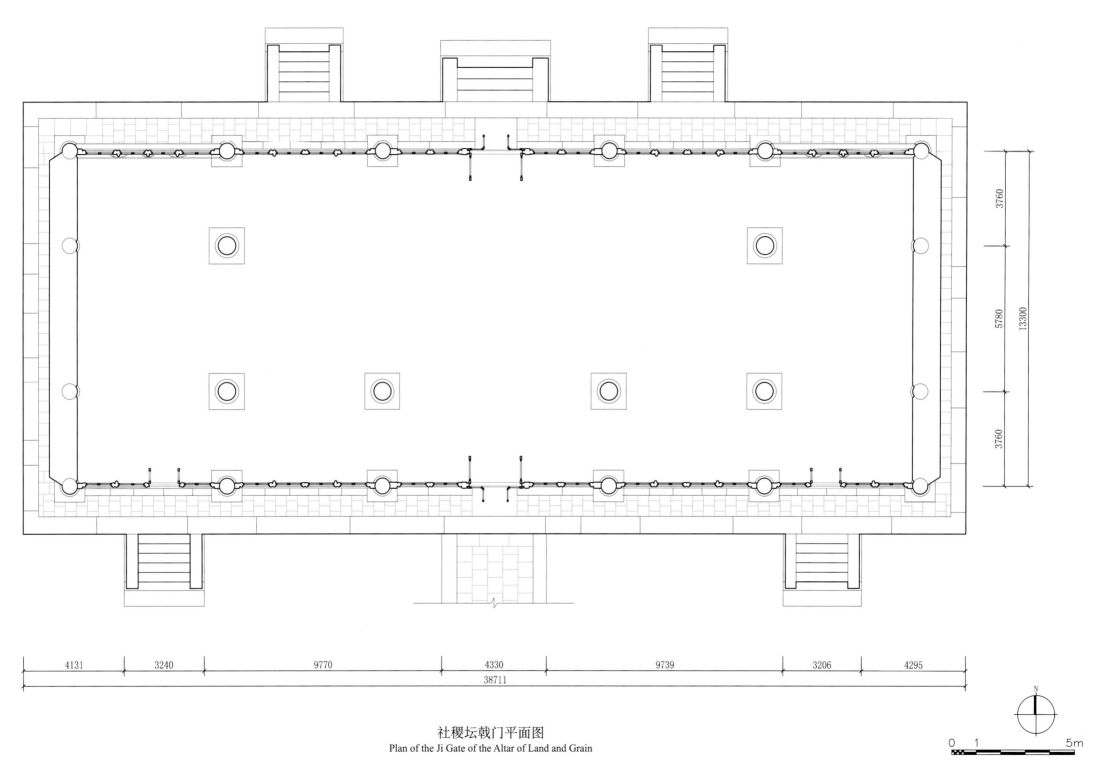

3760
5780
13300
3760

4131　3240　9770　4330　9739　3206　4295
38711

社稷坛戟门平面图
Plan of the Ji Gate of the Altar of Land and Grain

N

0　1　5m

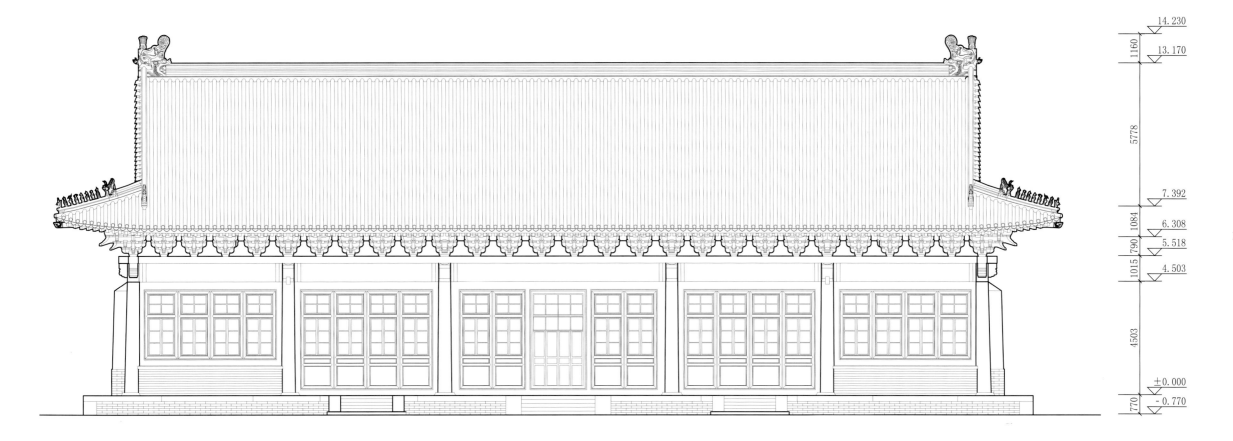

14.230
13.170
1160
5778
7.392
1084
6.308
790
5.518
1015
4.503
4503
±0.000
770
-0.770

1183 | 1103 | 1103 | 6170 | 6400 | 9293 | 6170 | 6400 | 1103 | 1103 | 1183
41211

社稷坛戟门北立面图
North elevation of the Ji Gate of the Altar of Land and Grain

0  1          5m

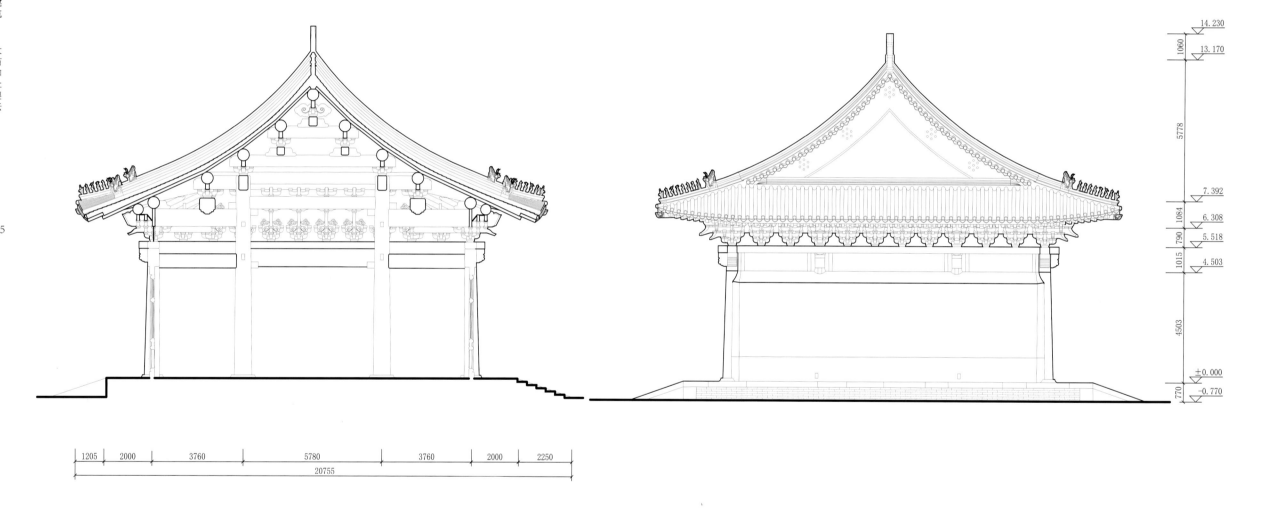

14.230
13.170
1060
5778
7.392
1084
6.308
790
5.518
1015
4.503
4503
±0.000
770
-0.770

1205 2000 3760 5780 3760 2000 2250
20755

社稷坛戟门明间剖面图
Section of the central bay of the Ji Gate of the Altar of Land and Grain

社稷坛戟门东立面图
East elevation of the Ji Gate of the Altar of Land and Grain

0 1 5m

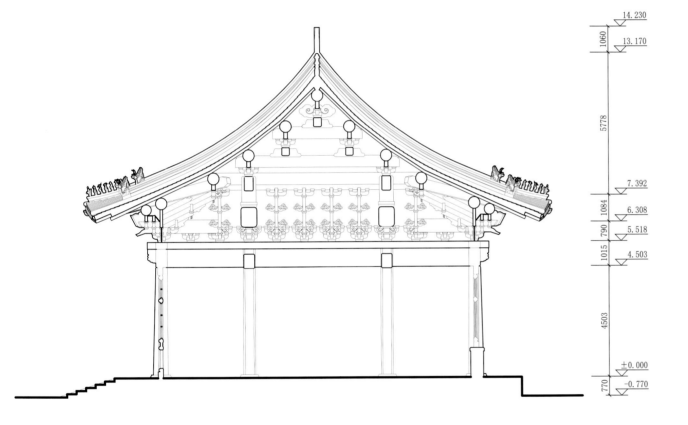

14.230

1060

13.170

5778

7.392

1084

6.308

790

5.518

1015

4.503

4503

±0.000

770

-0.770

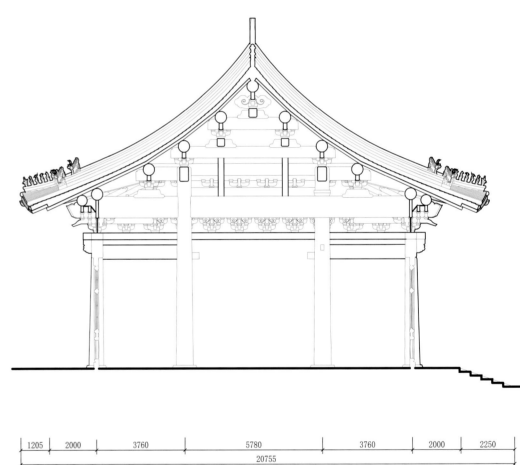

| 1205 | 2000 | 3760 | 5780 | 3760 | 2000 | 2250 |

20755

社稷坛戟门梢间剖面图
Section of the terminal bay of the Ji Gate of the Altar of Land and Grain

社稷坛戟门次间剖面图
Section of the side bay of the Ji Gate of the Altar of Land and Grain

0　1　　　　5m

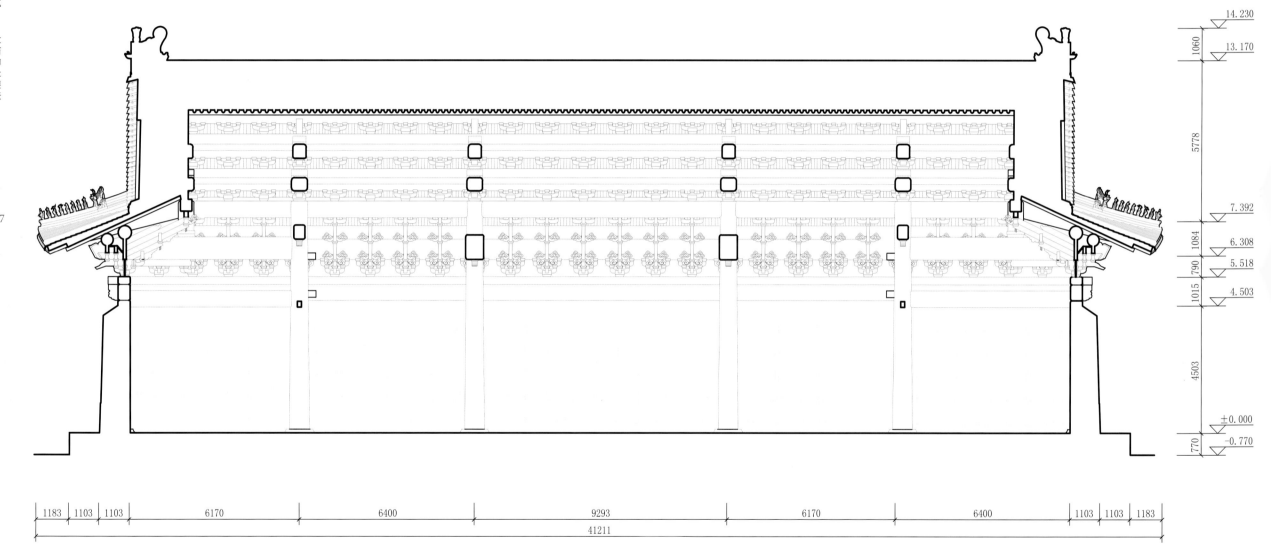

14.230
13.170
7.392
6.308
5.518
4.503
±0.000
-0.770

1060
5778
1084
790
1015
4503
770

1183 | 1103 | 1103 | 6170 | 6400 | 9293 | 6170 | 6400 | 1103 | 1103 | 1183

41211

社稷坛戟门纵剖面图
Longitudinal Section of the Ji Gate of the Altar of Land and Grain

0   1                     5m

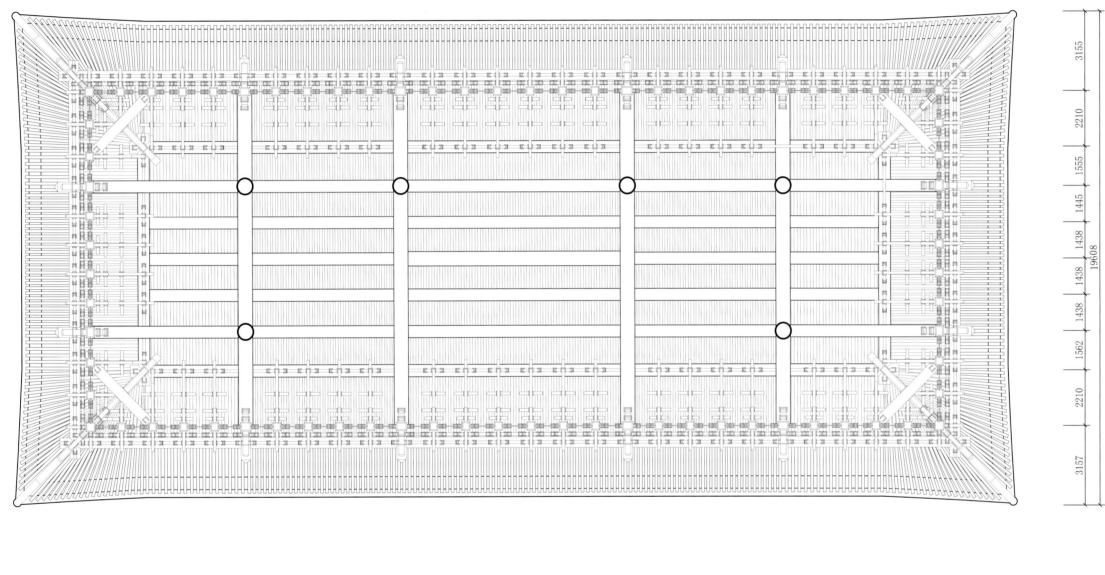

3155
2210
1555
1445
1438
1438
1438
1562
2210
3157

19608

3163 2208 4189 6400 9323 6400 4189 2208 3162

41242

社稷坛戟门天花仰视图
Top view of the ceiling of the Ji Gate of the Altar of Land and Grain

0 1 5m

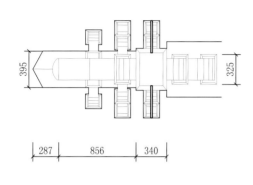

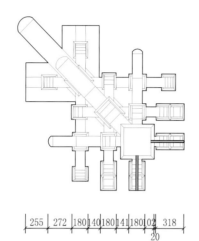

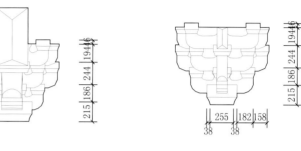

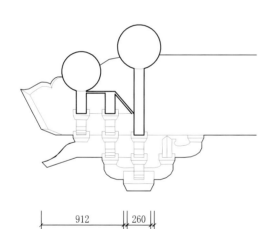

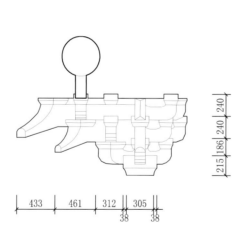

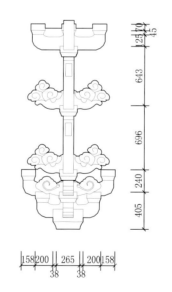

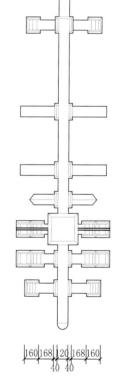

社稷坛戟门柱头科及角科斗栱大样图

Detail of the connecting bracket sets on columns and corner columns of the Ji Gate of the Altar of Land and Grain

0    0.4    0.8m

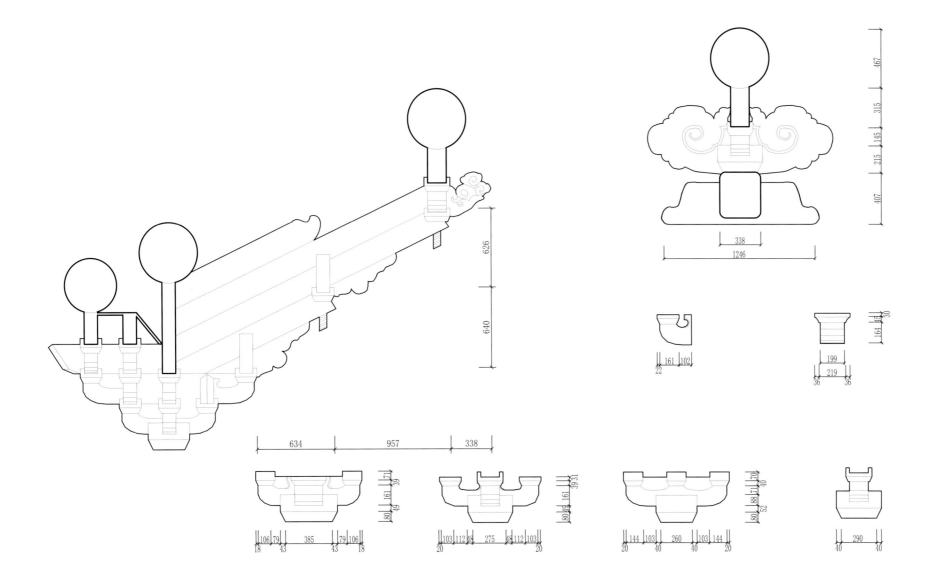

社稷坛戟门镏金斗栱平身科大样
Detail of the cantilevered bracket sets between columns of the Ji Gate of the Altar of Land and Grain

0    0.3    0.6m

# 大殿
## The Main Hall

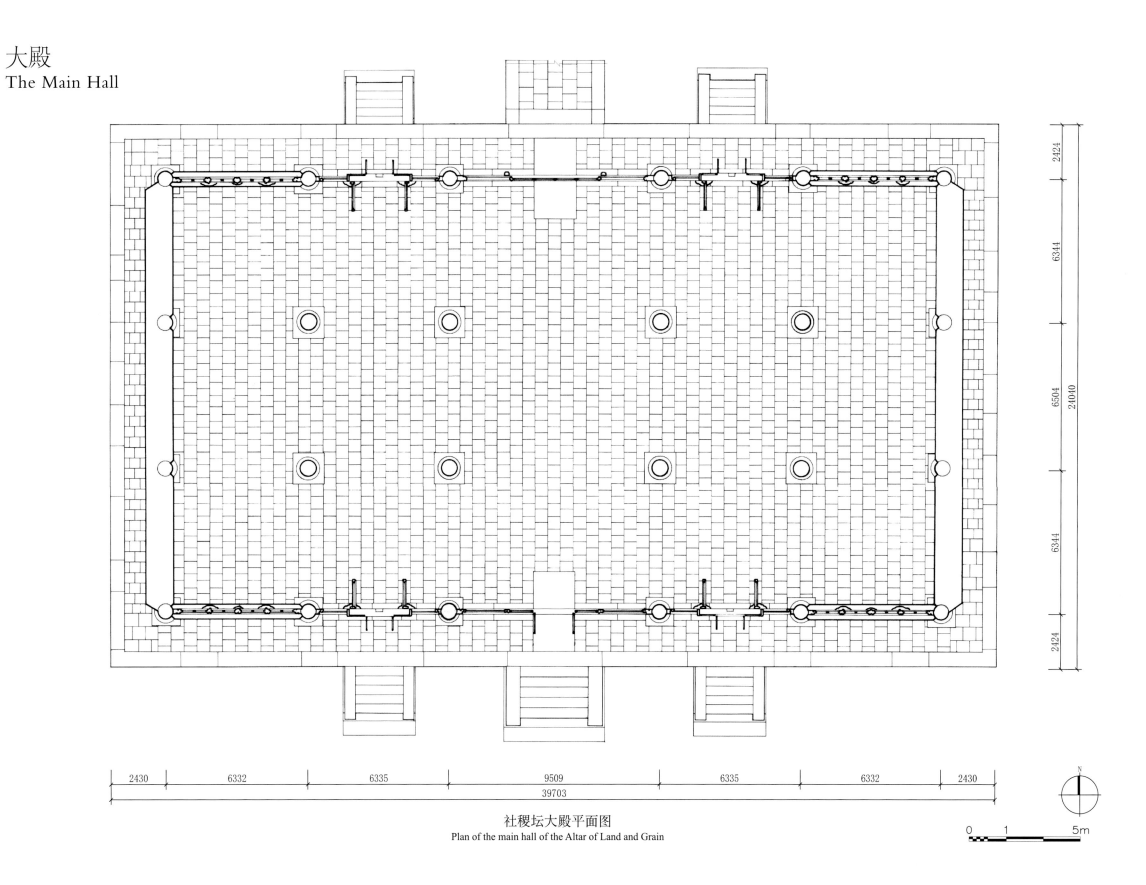

社稷坛大殿平面图
Plan of the main hall of the Altar of Land and Grain

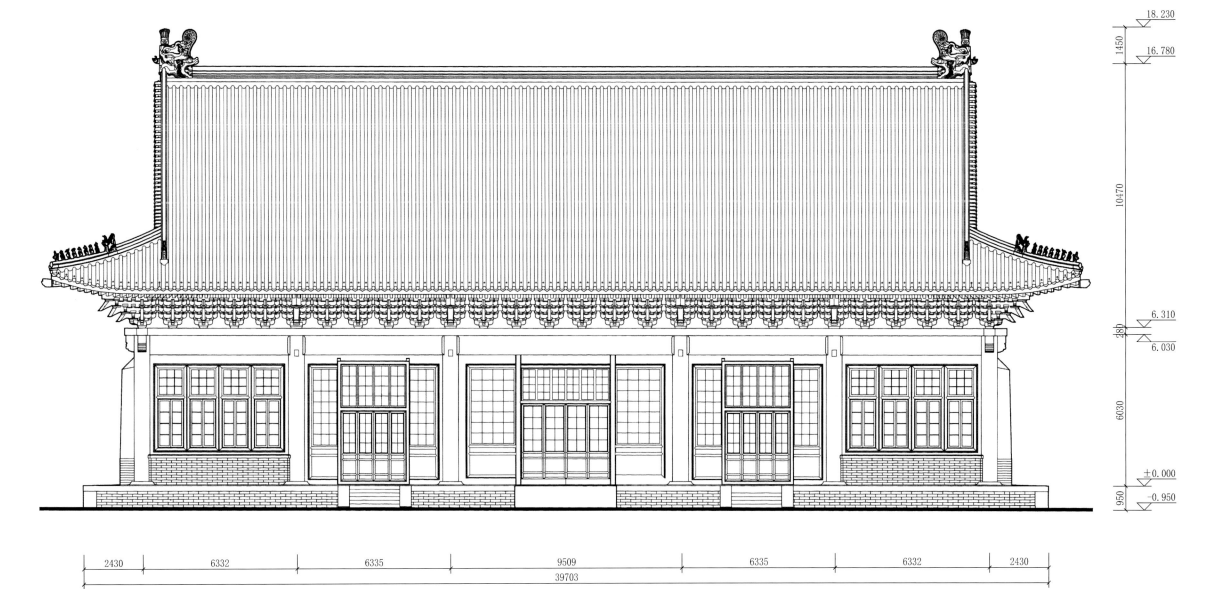

18.230

1450

16.780

10470

6.310

380

6.030

6030

±0.000

950

-0.950

2430    6332    6335    9509    6335    6332    2430

39703

社稷坛大殿北立面图
North elevation of the main hall of the Altar of Land and Grain

0    1    5m

18.230

8230

10.000

3690

6.310

280

6.030

6030

±0.000

950

-0.950

社稷坛大殿东立面图
East elevations of the main hall of the Altar of Land and Grain

0    1                    5m

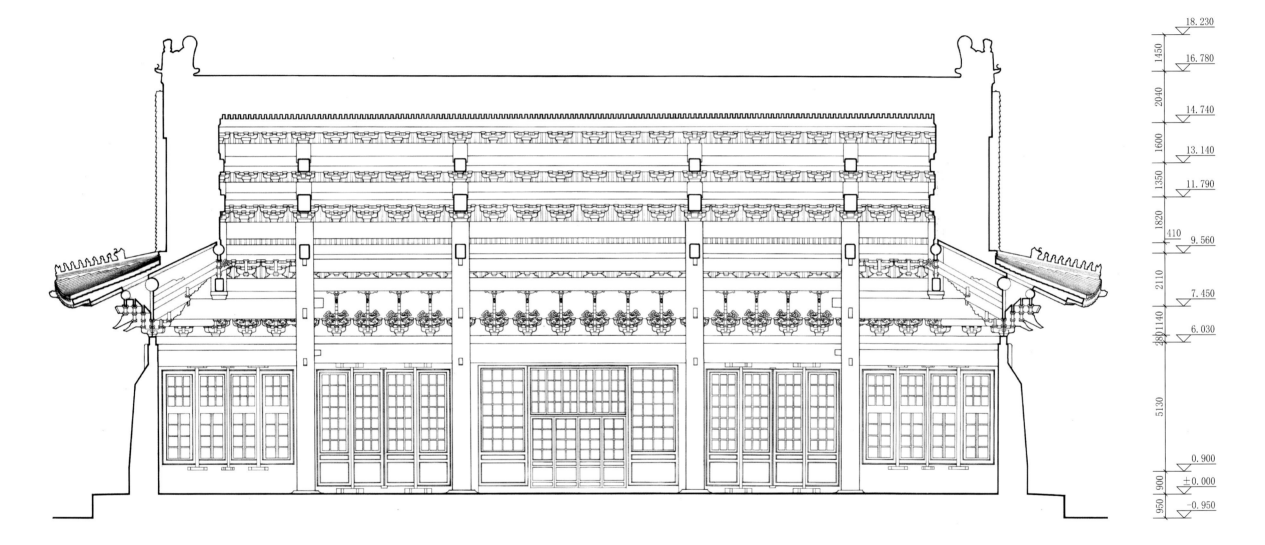

社稷坛大殿纵剖面图
Longitudinal section of the main hall of the Altar of Land and Grain

0　1　　　　5m

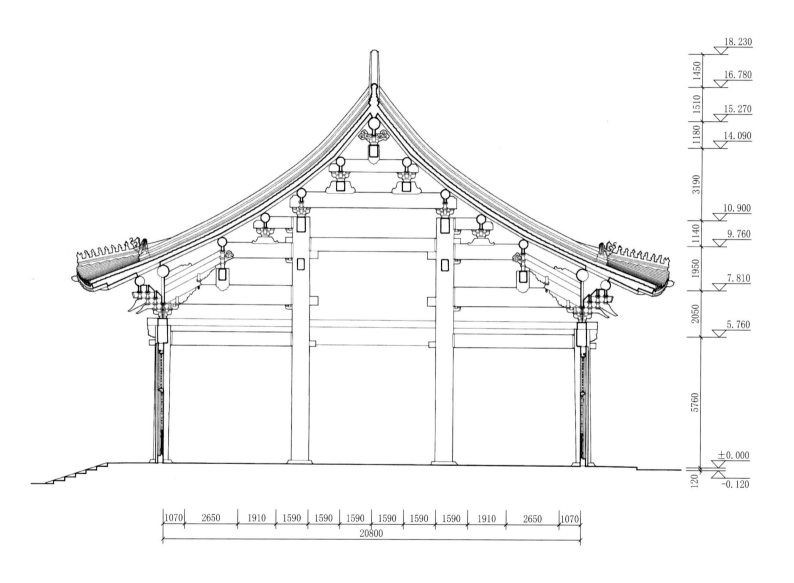

18.230

1450

16.780

1510

15.270

1180

14.090

3190

10.900

1140

9.760

1950

7.810

2050

5.760

5760

±0.000

120

-0.120

| 1070 | 2650 | 1910 | 1590 | 1590 | 1590 | 1590 | 1590 | 1590 | 1910 | 2650 | 1070 |

20800

社稷坛大殿明间剖面图
Sections of the central bay of the main hall of the Altar of Land and Grain

0   1        5m

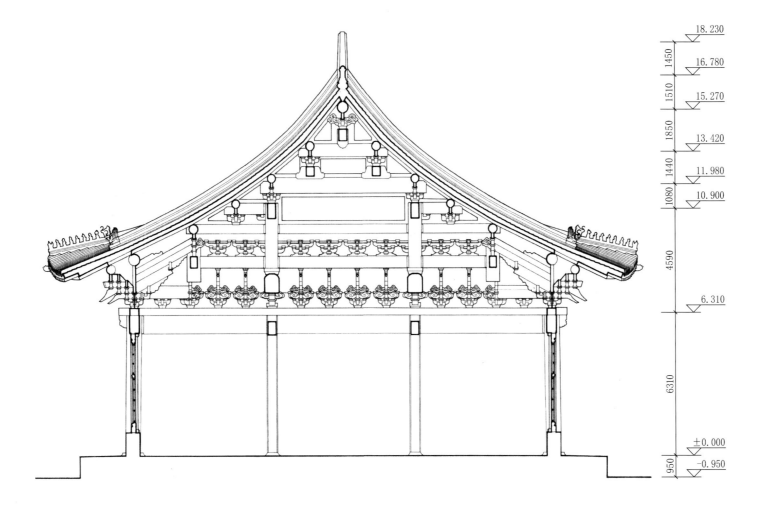

18.230
1450
16.780
1510
15.270
1850
13.420
1440
11.980
1080
10.900

4590

6.310

6310

±0.000
950
-0.950

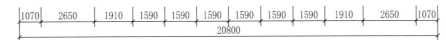

| 1070 | 2650 | 1910 | 1590 | 1590 | 1590 | 1590 | 1590 | 1590 | 1910 | 2650 | 1070 |

20800

社稷坛大殿梢间剖面图
Sections of the terminal bay of the main hall of the Altar of Land and Grain

0    1         5m

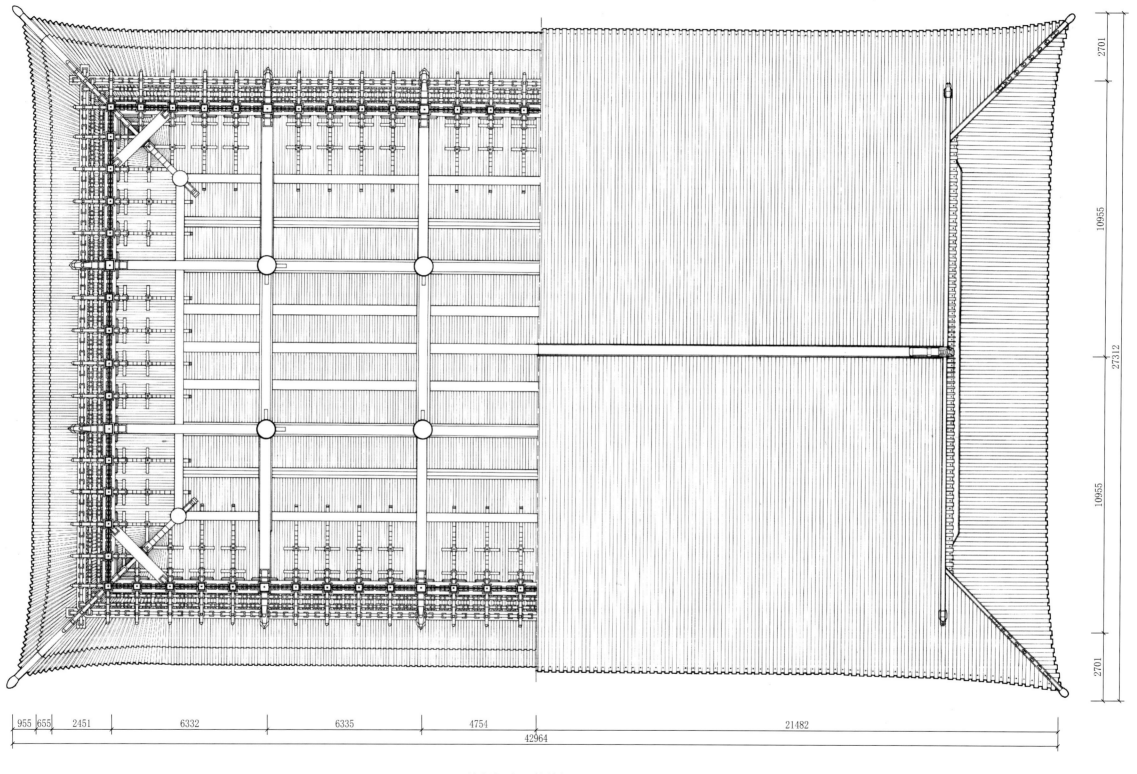

社稷坛大殿仰俯视平面图
Top and bottom plans of the main hall of the Altar of Land and Grain

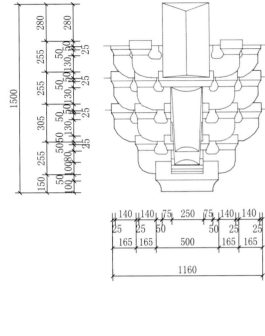

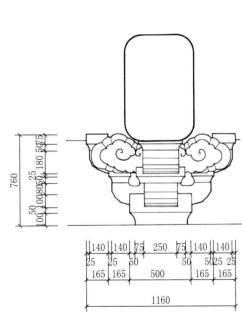

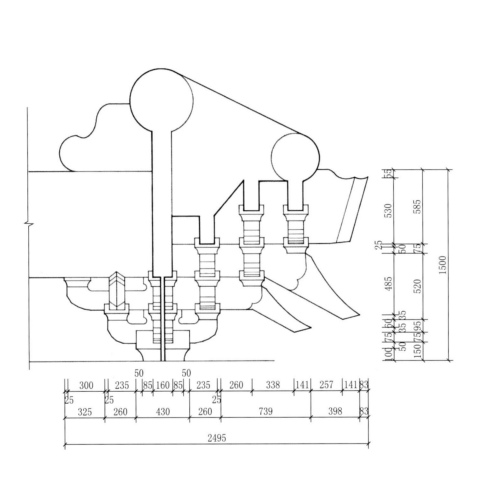

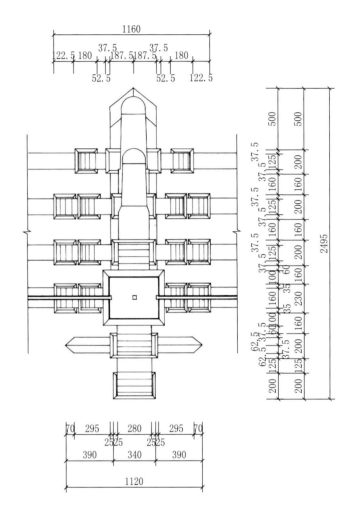

社稷坛大殿外檐柱头科斗栱大样

Detail of the connecting bracket sets on outer columns of the Main Hall of the Altar of Land and Grain

0　0.3　0.6m

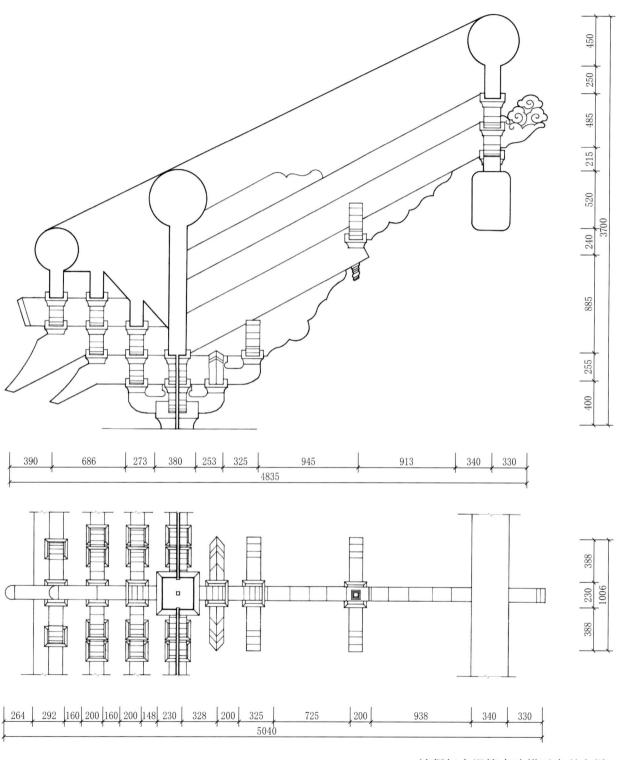

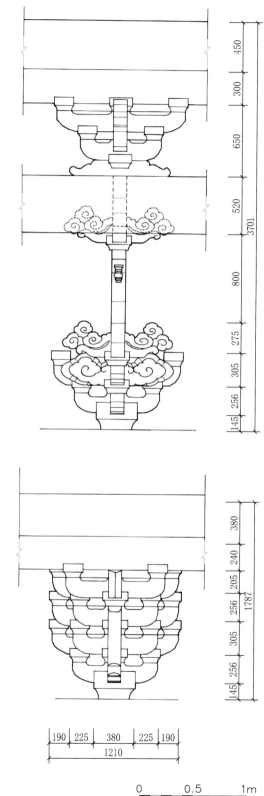

社稷坛大殿镏金斗栱平身科大样

Detail of the cantilevered bracket sets between columns of the main hall of the Altar of Land and Grain

0　　　0.5　　　1m

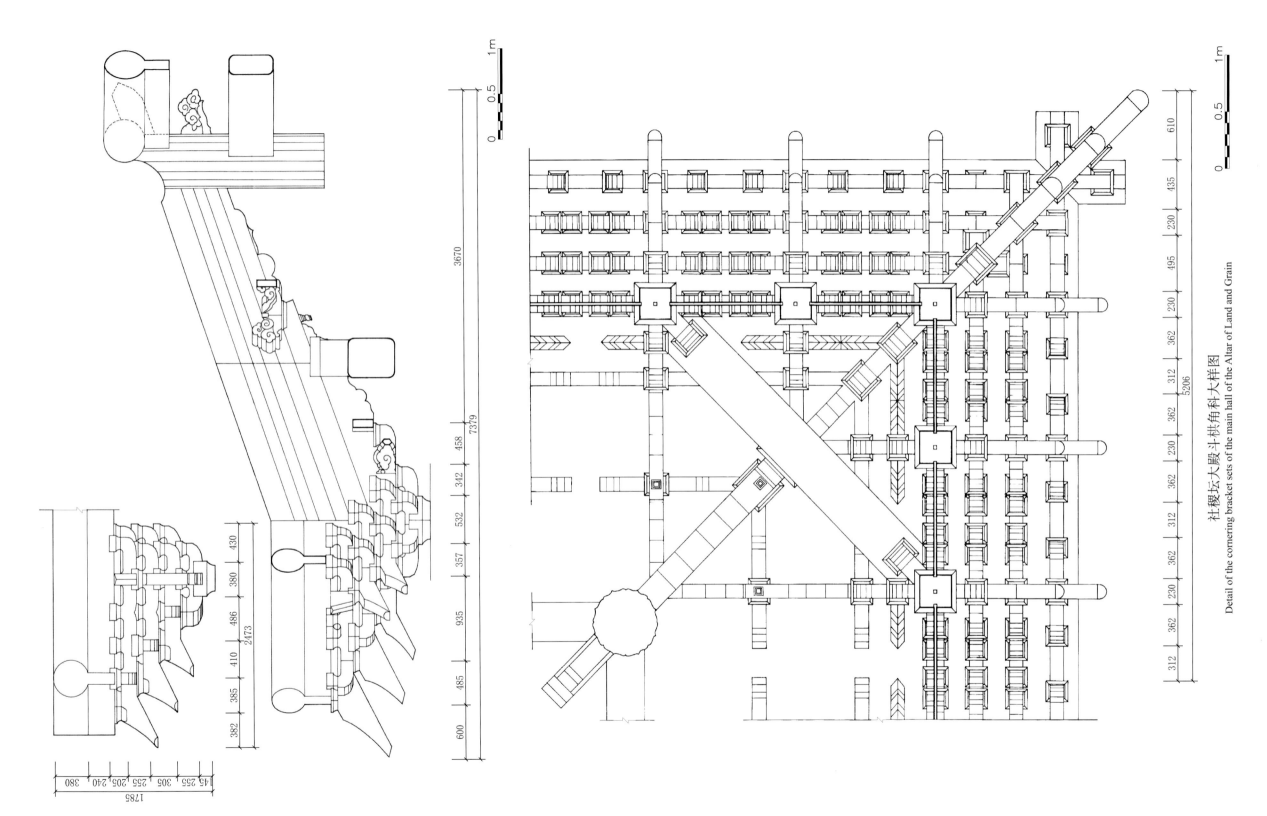

社稷坛大殿斗栱角科大样图

Detail of the cornering bracket sets of the main hall of the Altar of Land and Grain

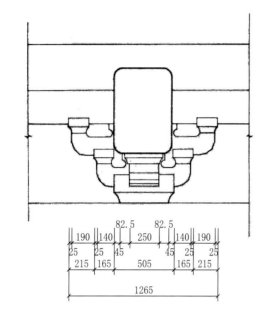

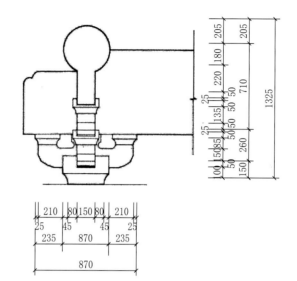

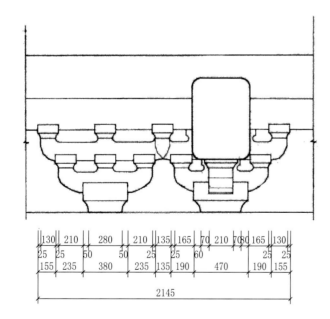

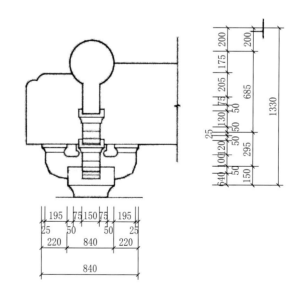

社稷坛大殿内檐柱头科斗栱大样图

Detail of the connecting bracket sets on inner columns of the Main Hall of the Altar of Land and Grain

0    0.15    0.3m

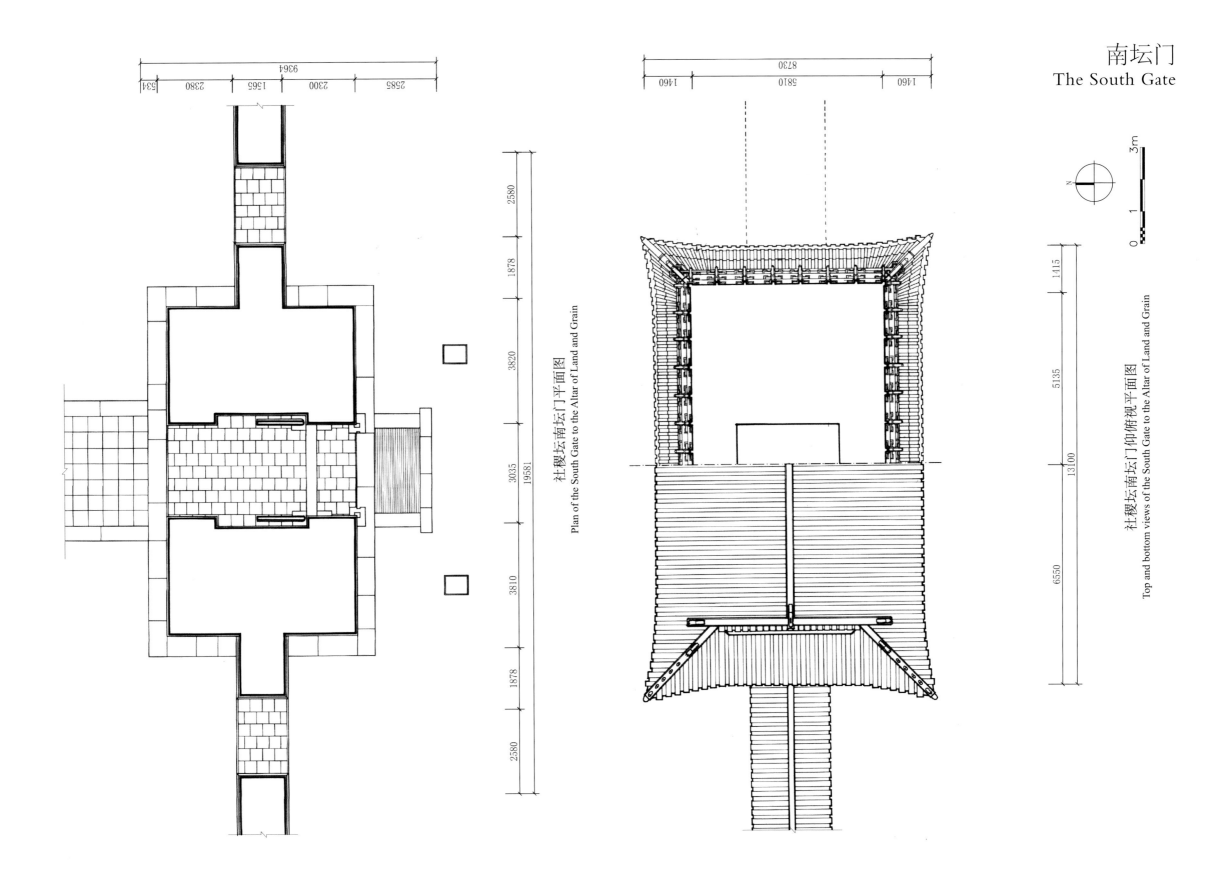

社稷坛南坛门平面图
Plan of the South Gate to the Altar of Land and Grain

社稷坛南坛门仰俯视平面图
Top and bottom views of the South Gate to the Altar of Land and Grain

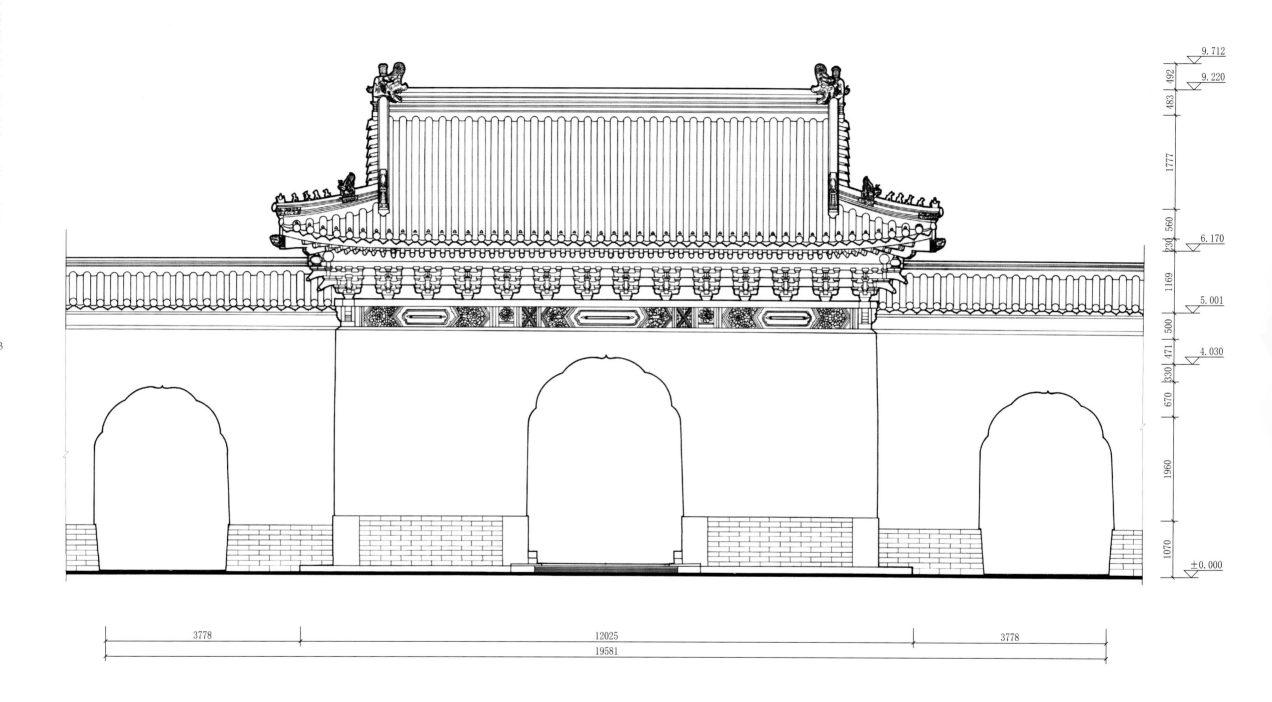

9.712

9.220

492
483

1777

560
230

6.170

1169

5.001

500

4.030

471
330

670

1960

1070

±0.000

3778

12025

3778

19581

社稷坛南坛门正立面图

Front elevation of the South Gate to the Altar of Land and Grain

0          1          2m

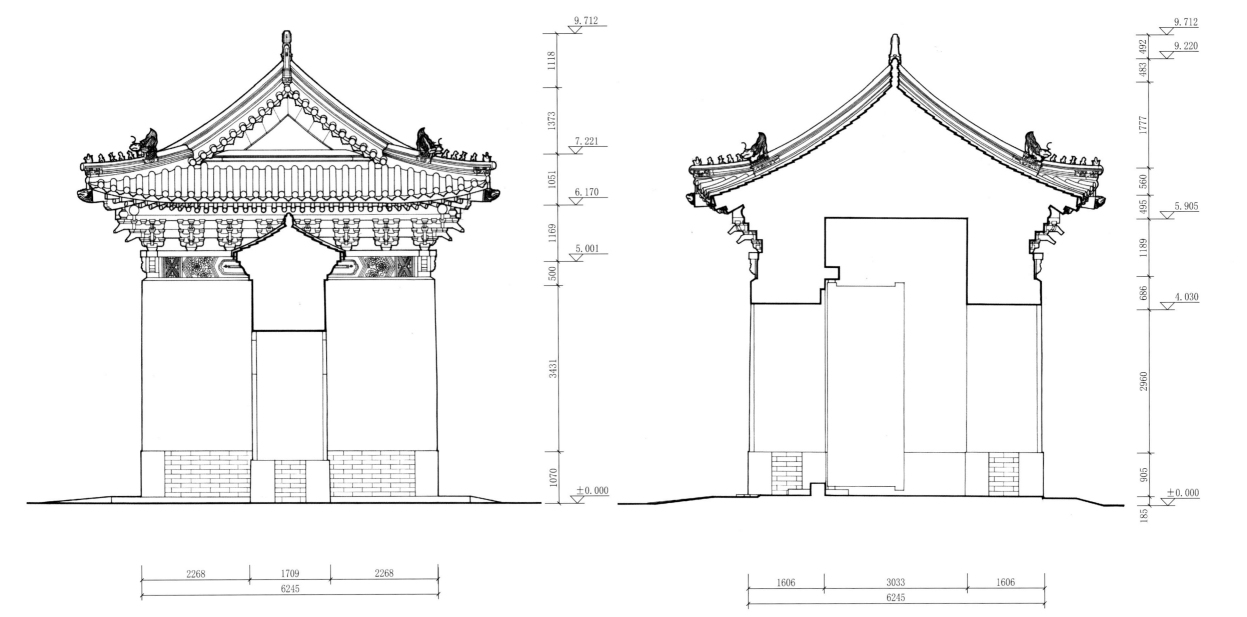

社稷坛南坛门侧立面图
Side elevation of the South Gate to the Altar of Land and Grain

社稷坛南坛门横剖面图
Section of the South Gate to the Altar of Land and Grain

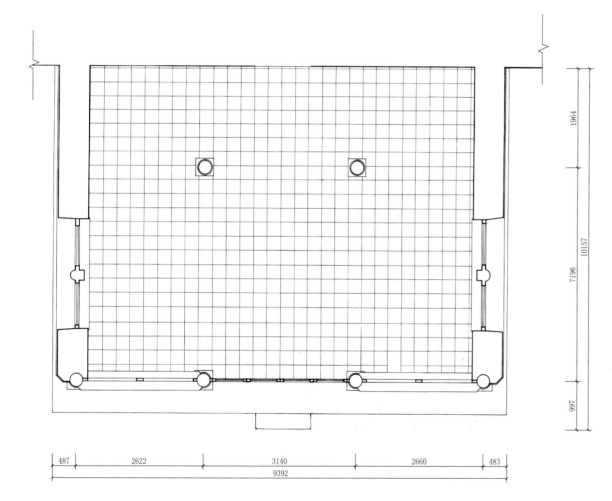

社稷坛南坛门值房平面图
Plan of the guardhouse of the South Gate of the Altar of Land and Grain

0    0.6    1.2m

社稷坛南坛门值房正立面图
Front elevation of the guardhouse of the South Gate to the Altar of Land and Grain

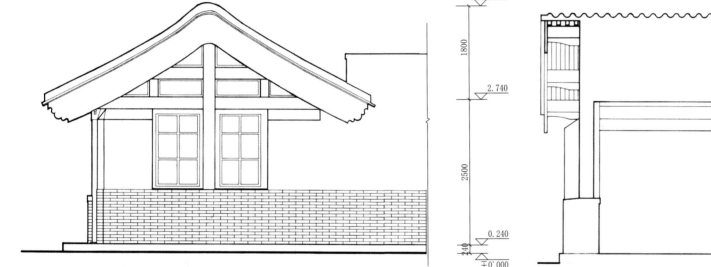

社稷坛南坛门值房侧立面图
Side elevation of the guardhouse of the South Gate to the Altar of Land and Grain

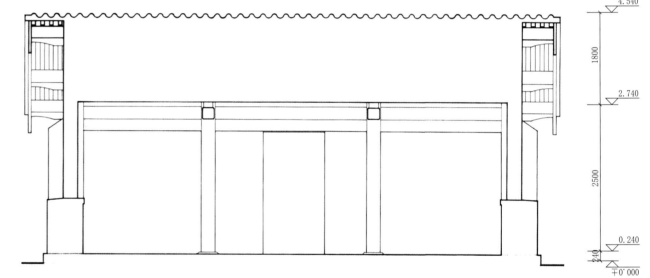

社稷坛南坛门值房横剖面图
Section of the guardhouse of the South Gate of the Altar of Land and Grain

0    0.6    1.2m

# 东坛门
## The East Gate

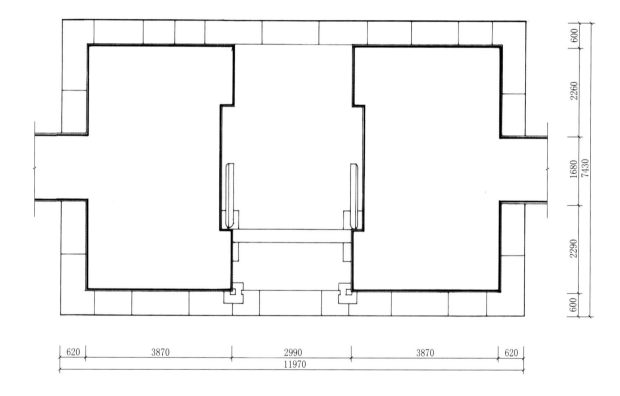

社稷坛东坛门平面图
Plan of the East Gate to the Altar of Land and Grain

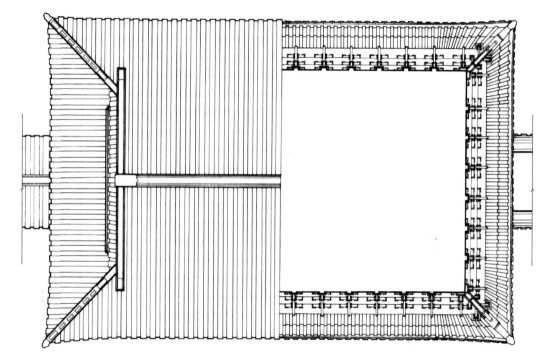

社稷坛东坛门屋顶平面及梁架仰视图
Roof plan and top view of the beam frame of the East Gate to the Altar of Land and Grain

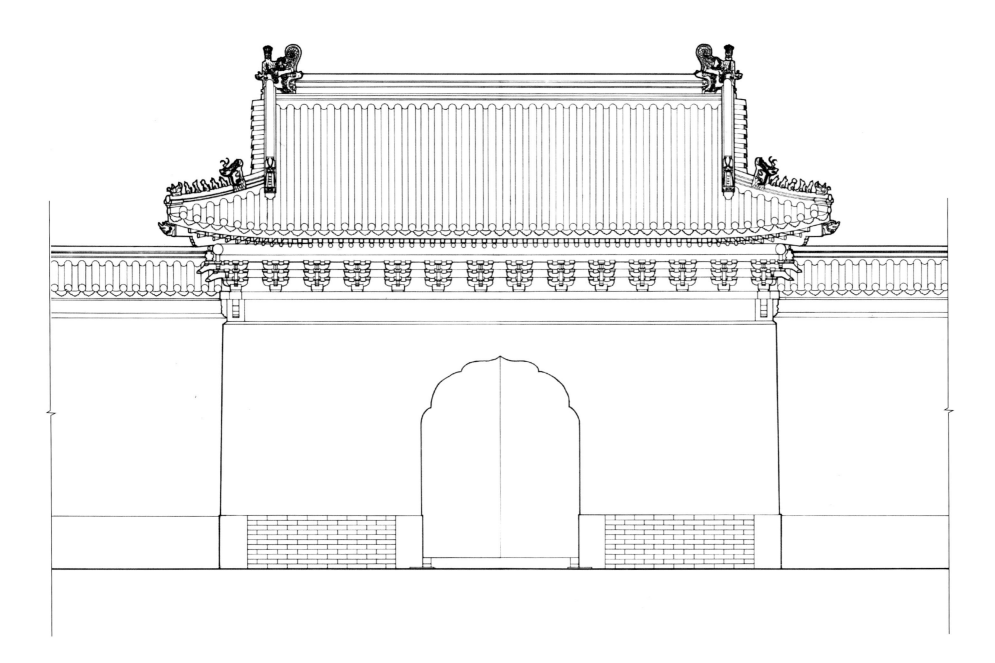
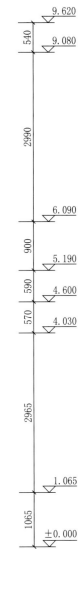

社稷坛东坛门正立面图
Front elevation of the East Gate to the Altar of Land and Grain

0    1    2m

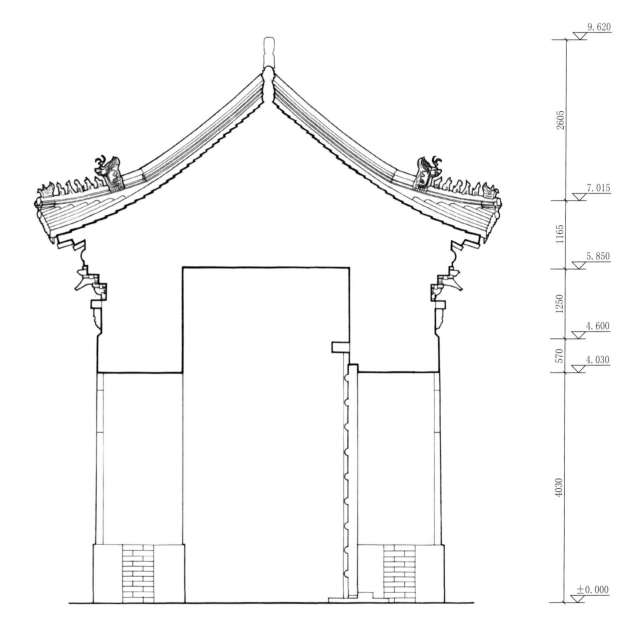

社稷坛东坛门侧立面图
Side elevation of the East Gate to the Altar of Land and Grain

社稷坛东坛门横剖面图
Section of the East Gate to the Altar of Land and Grain

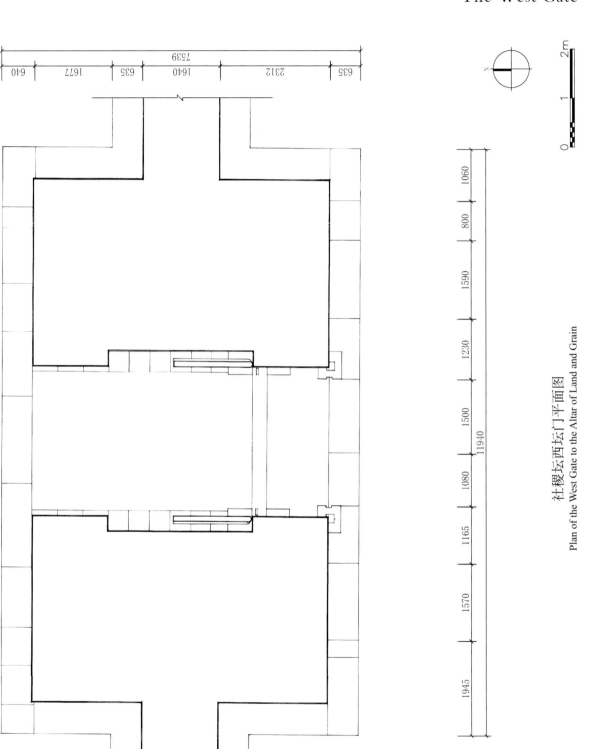

社稷坛西坛门屋顶平面及梁架仰视图
Roof plan and top view of the beam frame of the West Gate to the Altar of Land and Grain

社稷坛西坛门平面图
Plan of the West Gate to the Altar of Land and Grain

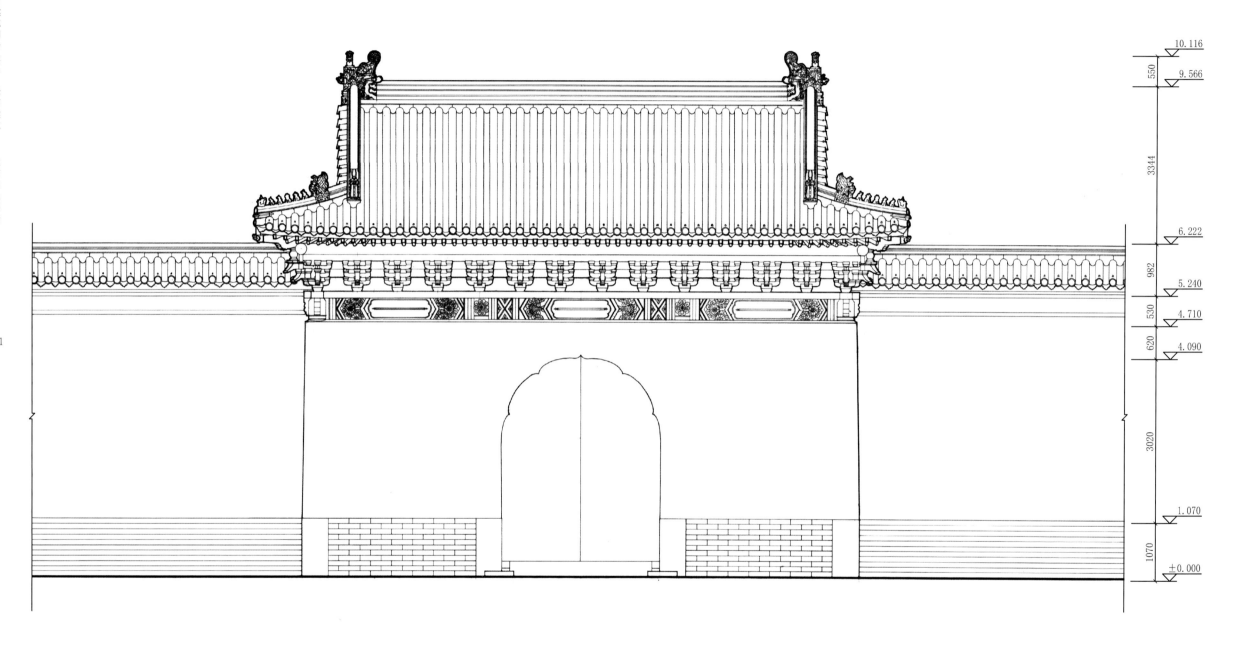

10.116

9.566

550

3344

6.222

982

5.240

530

4.710

620

4.090

3020

1.070

1070

±0.000

社稷坛西坛门正立面图

Front elevation of the West Gate to the Altar of Land and Grain

0    1    2m

6.922
2832
4.090
3020
1.070
1070
±0.000

1622　3020　1622
6264

社稷坛西坛门侧立面图
Side elevation of the West Gate to the Altar of Land and Grain

2312　1640　2312
6264

社稷坛西坛门横剖面图
Section of the West Gate to the Altar of Land and Grain

0　1　2m

# 左门
## The Left Gate

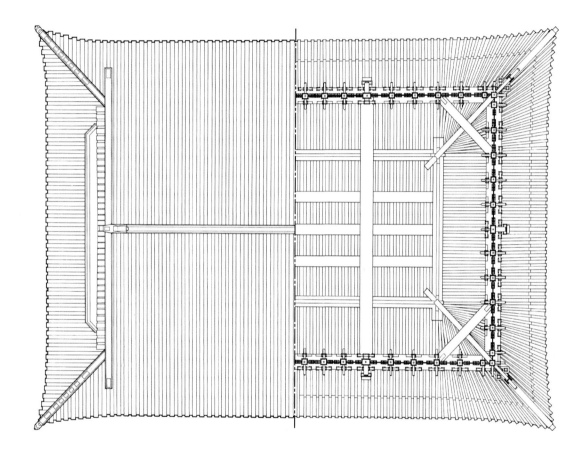

社稷坛左门屋顶平面及梁架仰视图

Roof plan and top view of the beam frame of the Left Gate to the Altar of Land and Grain

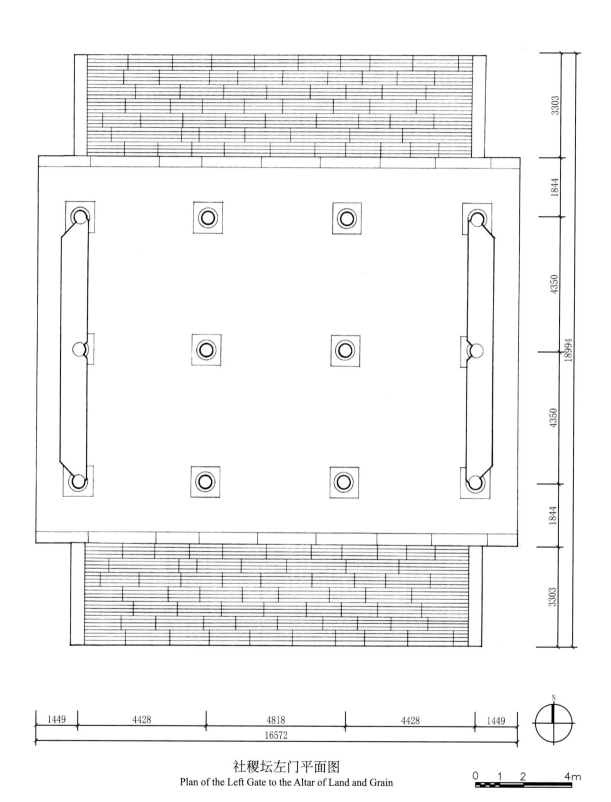

社稷坛左门平面图

Plan of the Left Gate to the Altar of Land and Grain

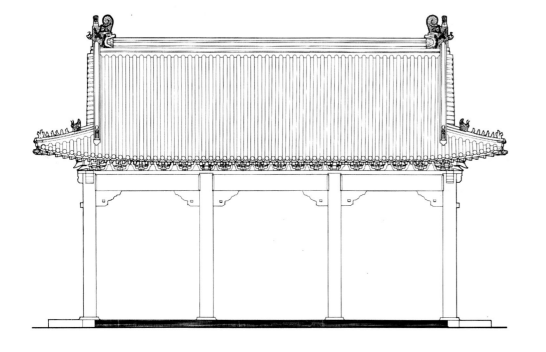

社稷坛左门正立面图
Front elevation of the Left Gate to the Altar of Land and Grain

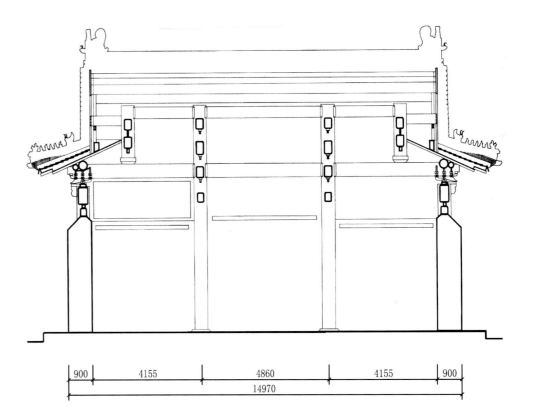

| 900 | 4155 | 4860 | 4155 | 900 |

14970

社稷坛左门纵剖面图
Longitudinal section of the Left Gate to the Altar of Land and Grain

0  1                    4m

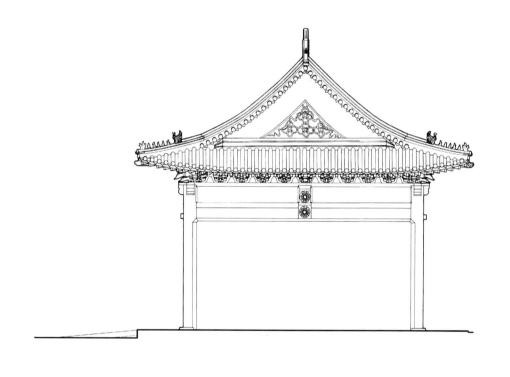

11.130

3920

7.210

1500

5.710

1620

4.090

4090

±0.000

社稷坛左门侧立面图
Side elevation of the Left Gate to the Altar of Land and Grain

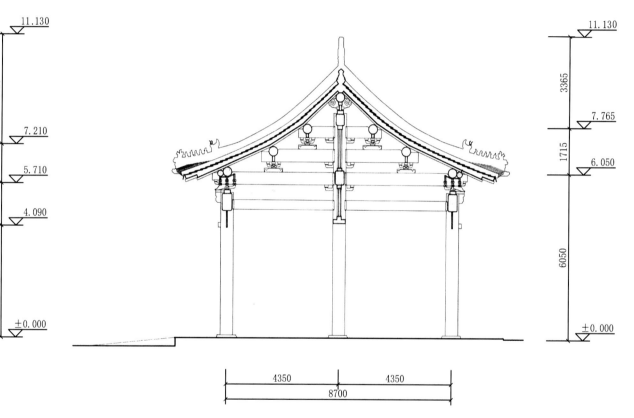

11.130

3365

7.765

1715

6.050

6050

±0.000

4350

4350

8700

社稷坛左门横剖面图
Section of the Left Gate to the Altar of Land and Grain

0　1　4m

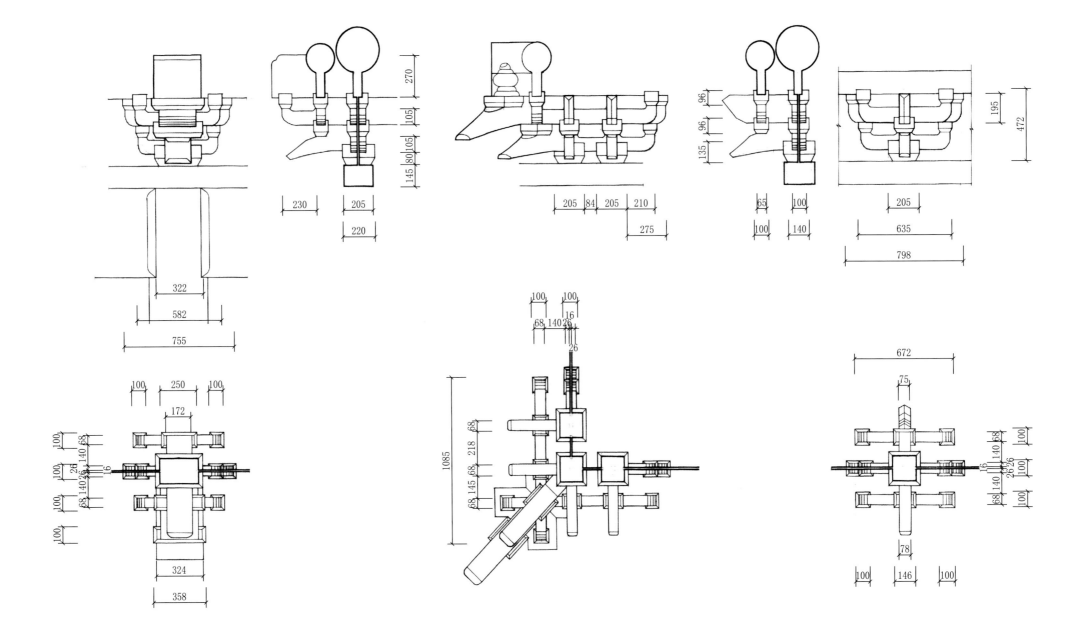

社稷坛左门斗栱大样
Detail of the connecting bracket sets of the Left Gate to the Altar of Land and Grain

0    0.25    0.5m

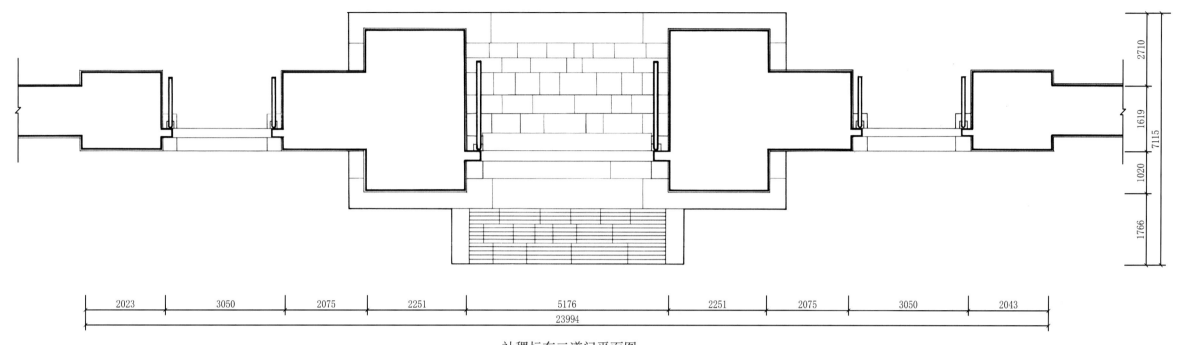

2023 3050 2075 2251 5176 2251 2075 3050 2043

23994

2710 1619 7115 1020 1766

### 社稷坛东二道门平面图
Plan of the East Second Gate to the Altar of Land and Grain

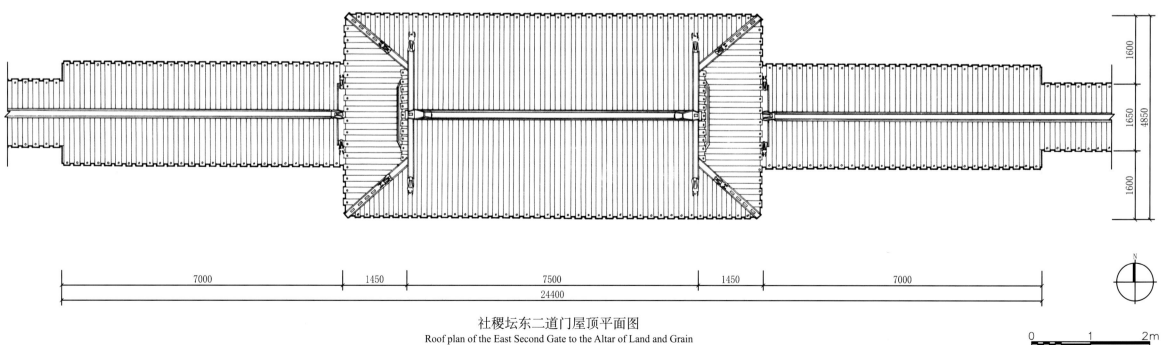

7000 1450 7500 1450 7000

24400

1600 1650 4850 1600

### 社稷坛东二道门屋顶平面图
Roof plan of the East Second Gate to the Altar of Land and Grain

0 1 2m

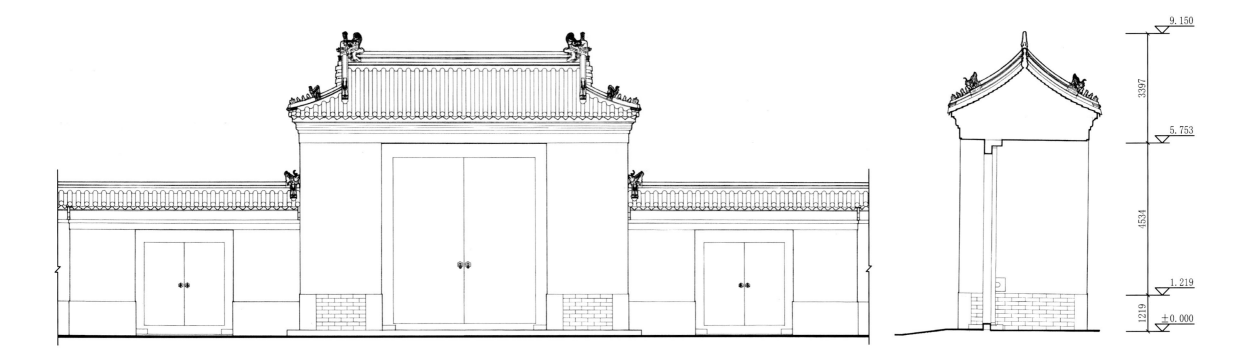

社稷坛东二道门正立面图
Front elevation of the East Second Gate to the Altar of Land and Grain

社稷坛东二道门剖面图
Section of the East Second Gate to the Altar of Land and Grain

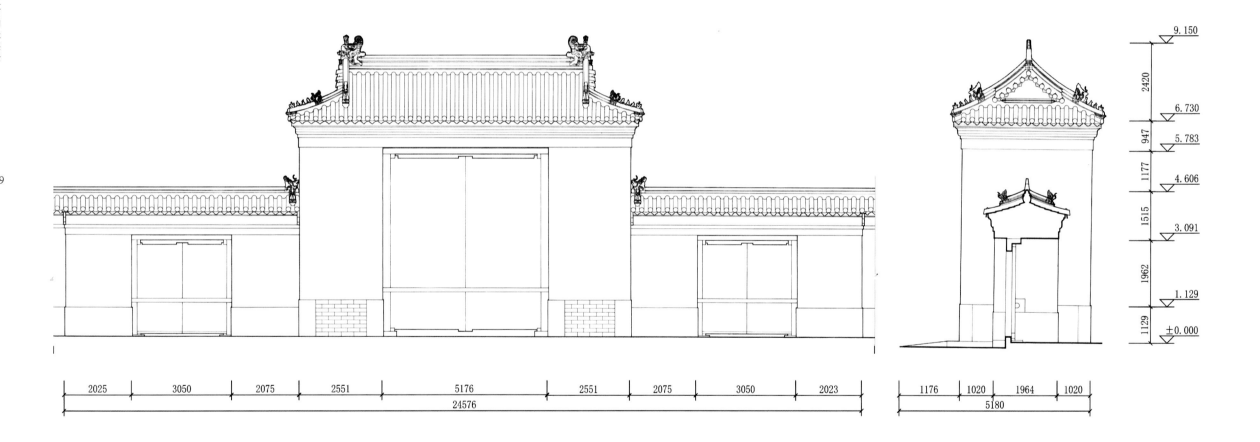

社稷坛东二道门背立面图
Rear elevation of the East Second Gate to the Altar of Land and Grain

社稷坛东二道门侧立面图
Side elevation of the East Second Gate to the Altar of Land and Grain

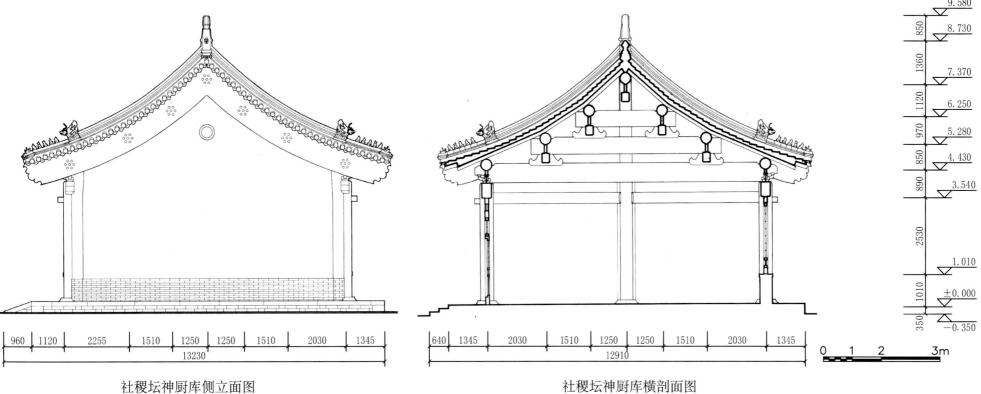

9.580
850
8.730
1360
7.370
1120
6.250
970
5.280
850
4.430
890
3.540
2530
1.010
1010
±0.000
350
−0.350

960 | 1120 | 2255 | 1510 | 1250 | 1250 | 1510 | 2030 | 1345
13230

640 | 1345 | 2030 | 1510 | 1250 | 1250 | 1510 | 2030 | 1345
12910

0 1 2 3m

社稷坛神厨库侧立面图
Side elevation of the Divine Kitchen Store of the Altar of Land and Grain

社稷坛神厨库横剖面图
Section of the Divine Kitchen Store of the Altar of Land and Grain

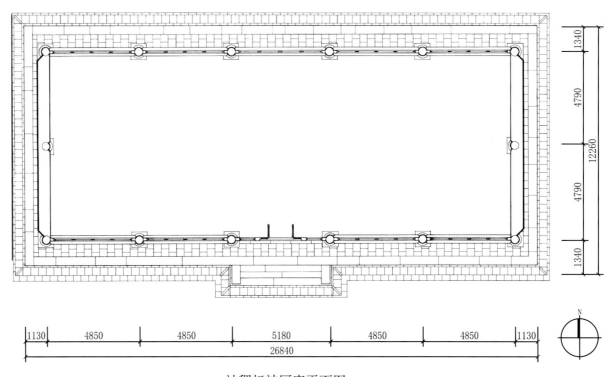

1340
4790
12260
4790
1340

1130 | 4850 | 4850 | 5180 | 4850 | 4850 | 1130
26840

N

0 1 4m

社稷坛神厨库平面图
Plan of the Divine Kitchen Store of the Altar of Land and Grain

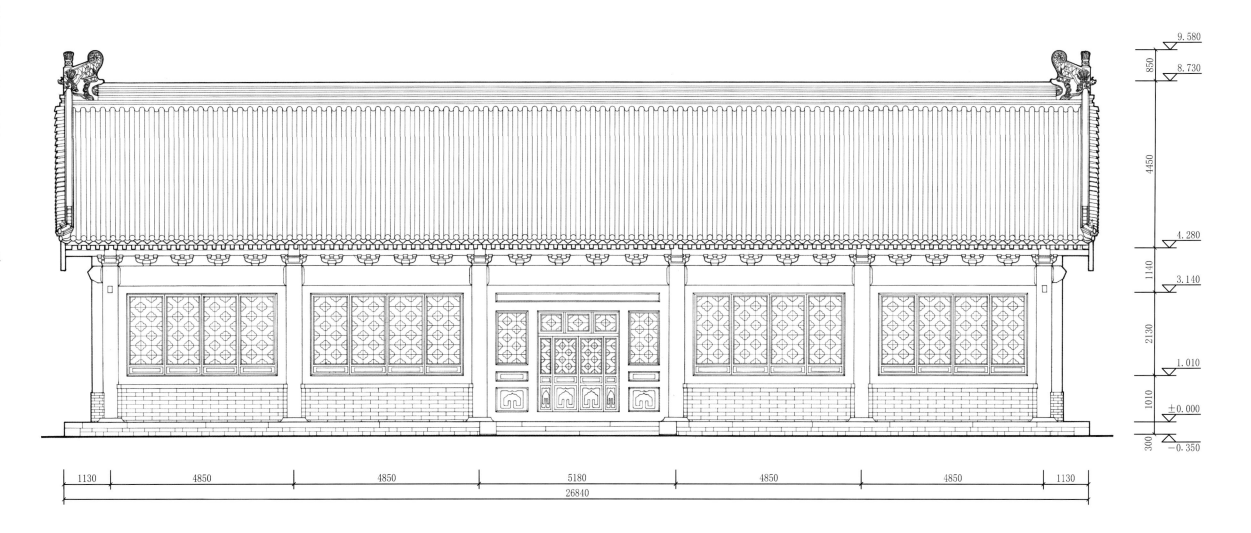

9.580
8.730
850
4450
4.280
1140
3.140
2130
1.010
1010
±0.000
300
−0.350

1130  4850  4850  5180  4850  4850  1130
26840

社稷坛神厨库正立面图
Front elevation of the Divine Kitchen Store of the Altar of Land and Grain

0  1  2m

9.580

850

8.730

2230

6.500

550 5.950

350 5.600

600 5.000

300 4.700

660 4.040

500 3.540

2530

1.010

1010 ±0.000

350 0.350

670 460　4850　4850　5180　4850　4850　460 670

26840

社稷坛神厨库纵剖面图
Section of the Divine Kitchen Store of the Altar of Land and Grain

0　1　2m

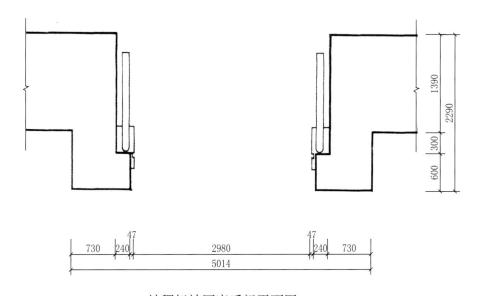

社稷坛神厨库后门平面图
Plan of the back door to the Divine Kitchen Store of the Altar of Land and Grain

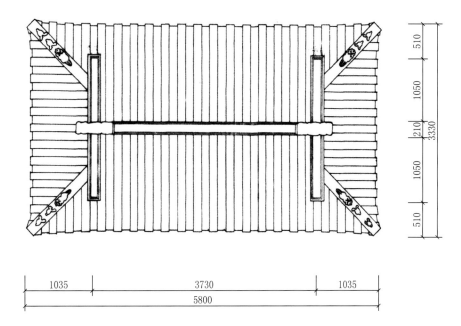

社稷坛神厨库后门屋顶平面图
Roof plan of the back door to the Divine Kitchen Store of the Altar of Land and Grain

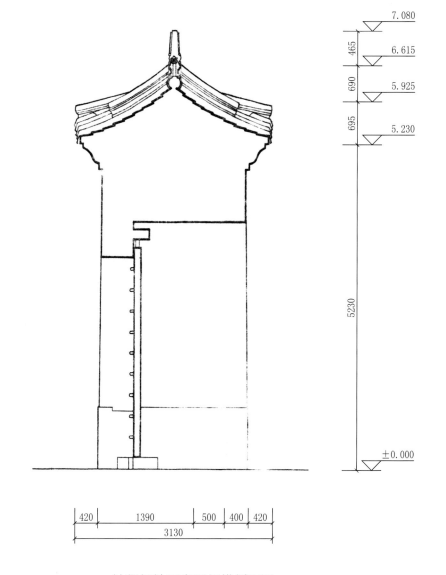

社稷坛神厨库后门横剖面图
Section of the back door to the Divine Kitchen Store of the Altar of Land and Grain

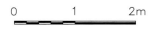

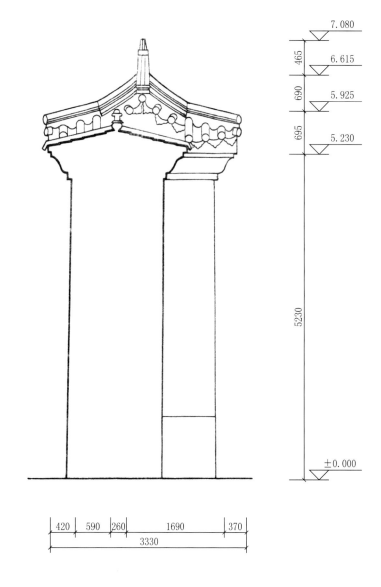

社稷坛神厨库后门正立面图
Front elevation of the back door to the Divine Kitchen Store of the Altar of Land and Grain

社稷坛神厨库后门侧立面图
Side elevation of the back door to the Divine Kitchen Store of the Altar of Land and Grain

0 1 2m

# 宰牲亭
The Butcher Pavilion

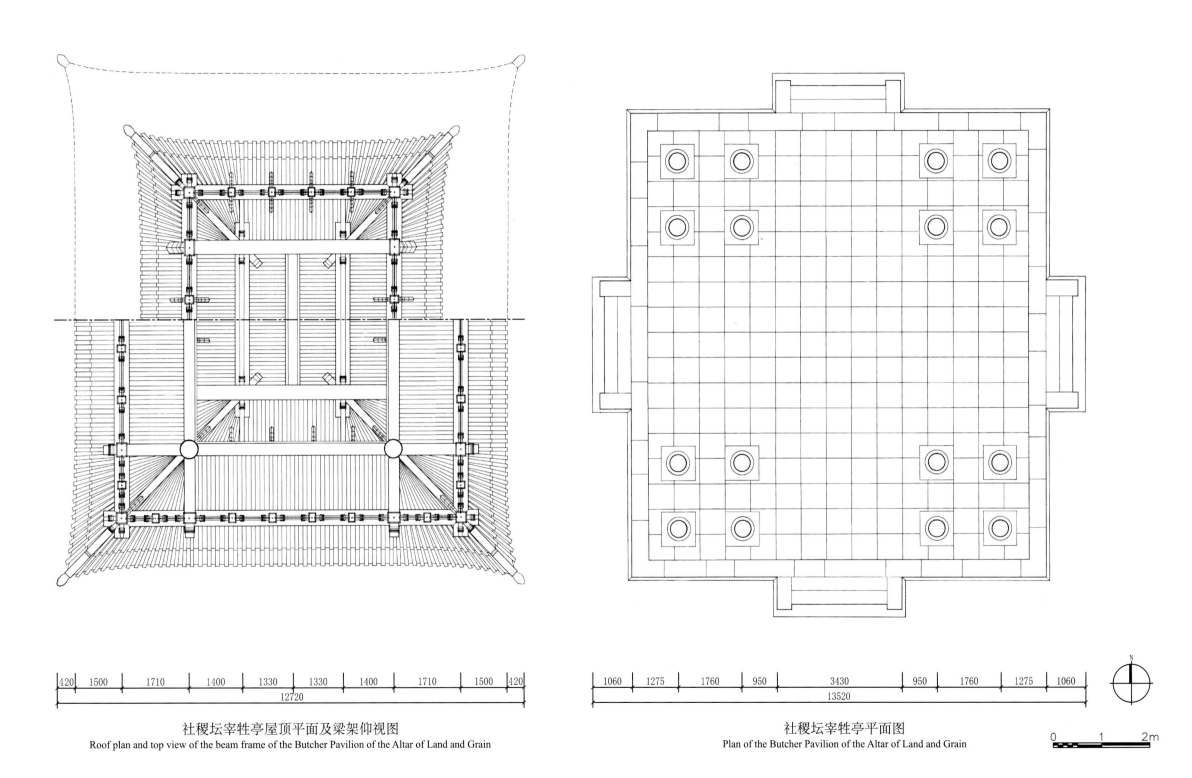

| 420 | 1500 | 1710 | 1400 | 1330 | 1330 | 1400 | 1710 | 1500 | 420 |
|---|---|---|---|---|---|---|---|---|---|

12720

| 1060 | 1275 | 1760 | 950 | 3430 | 950 | 1760 | 1275 | 1060 |
|---|---|---|---|---|---|---|---|---|

13520

社稷坛宰牲亭屋顶平面及梁架仰视图
Roof plan and top view of the beam frame of the Butcher Pavilion of the Altar of Land and Grain

社稷坛宰牲亭平面图
Plan of the Butcher Pavilion of the Altar of Land and Grain

0    1    2m

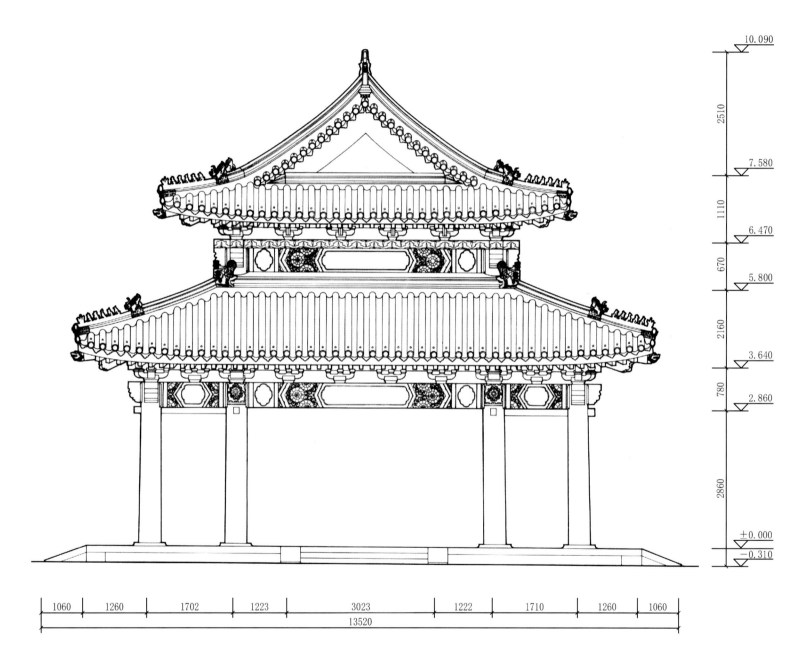

社稷坛宰牲亭北立面图
North elevation of the Butcher Pavilion of the Altar of Land and Grain

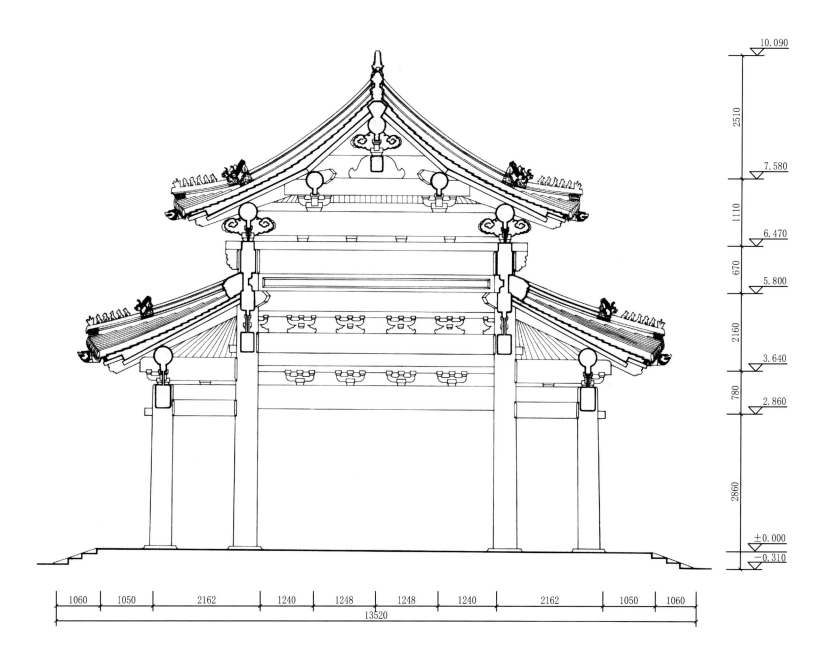

社稷坛宰牲亭横剖面图
Cross section of the Butcher Pavilion of the Altar of Land and Grain

0  1  2m

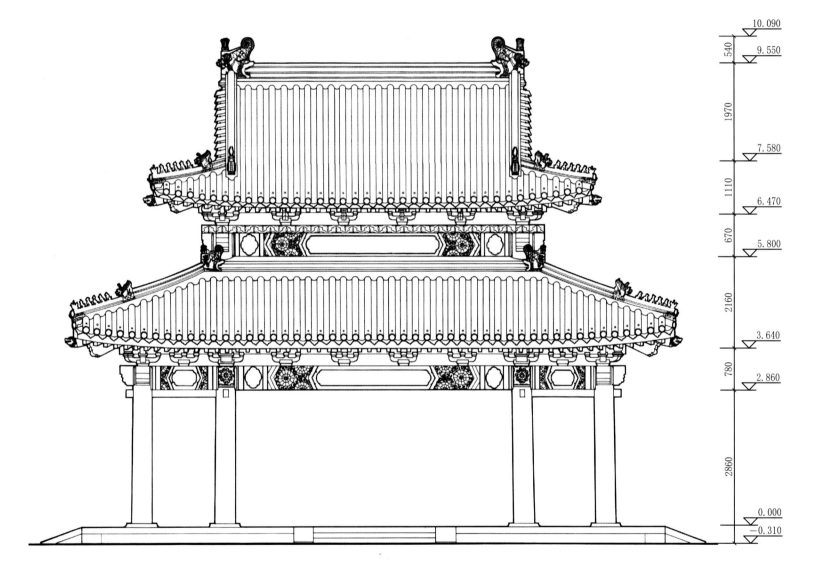

社稷坛宰牲亭东立面图
East elevation of the Butcher Pavilion of the Altar of Land and Grain

0 1 2m

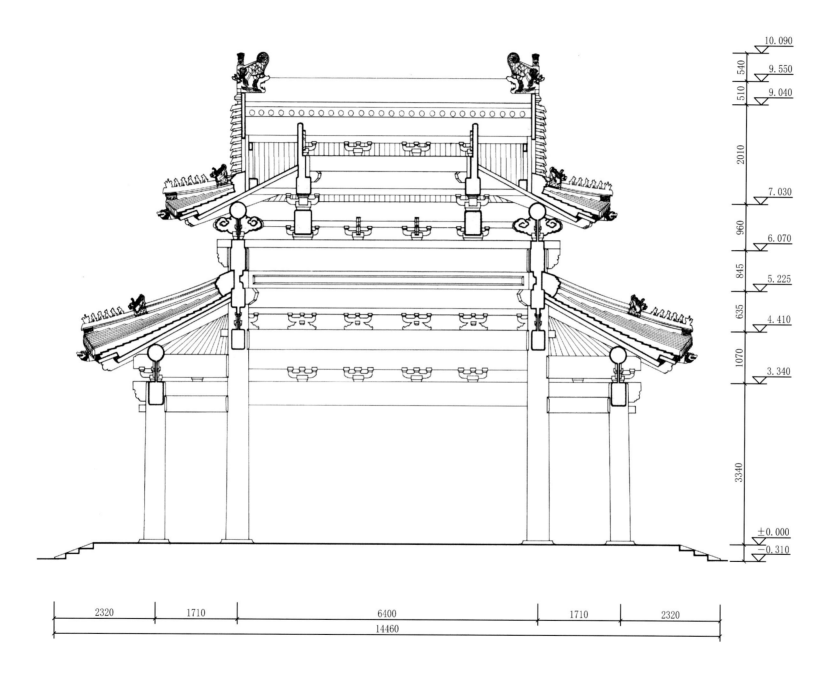

10.090
540 9.550
510 9.040
2010
7.030
960 6.070
845 5.225
635 4.410
1070 3.340
3340
±0.000
−0.310

2320　1710　6400　1710　2320
14460

社稷坛宰牲亭纵剖面图
Longitudinal section of the Butcher Pavilion of the Altar of Land and Grain

0　1　2m

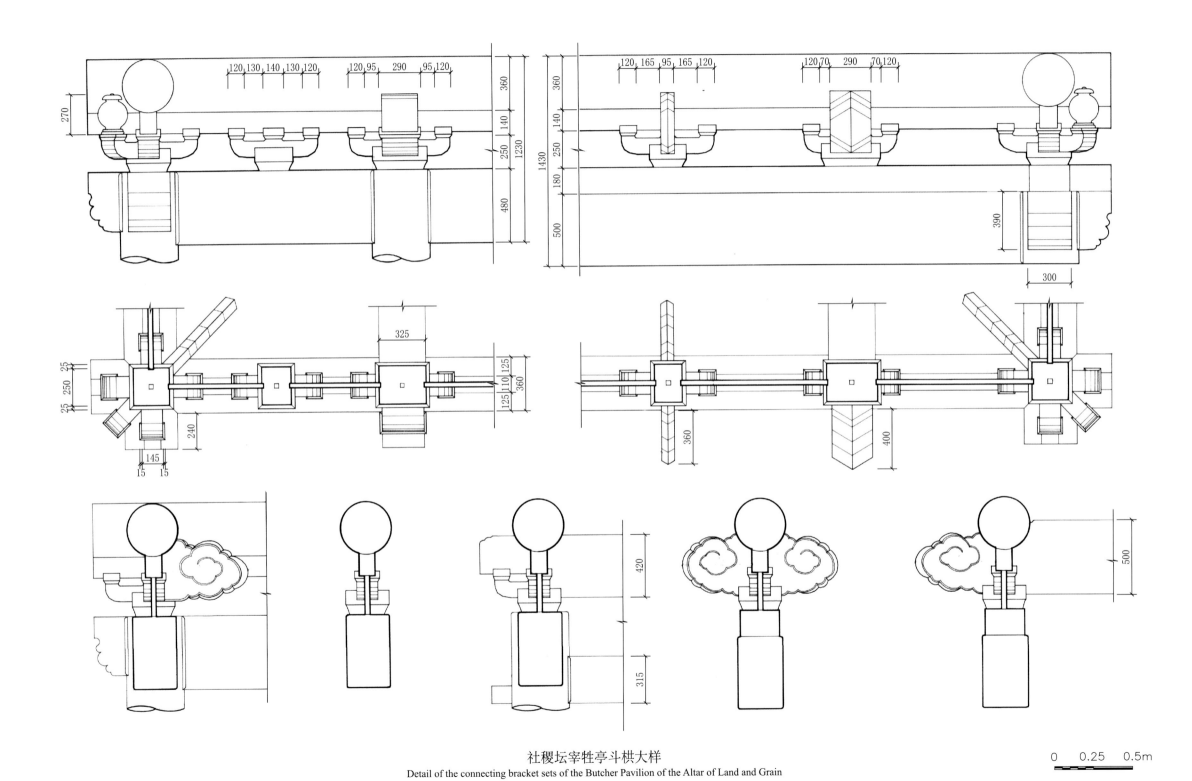

社稷坛宰牲亭斗栱大样
Detail of the connecting bracket sets of the Butcher Pavilion of the Altar of Land and Grain

0    0.25    0.5m

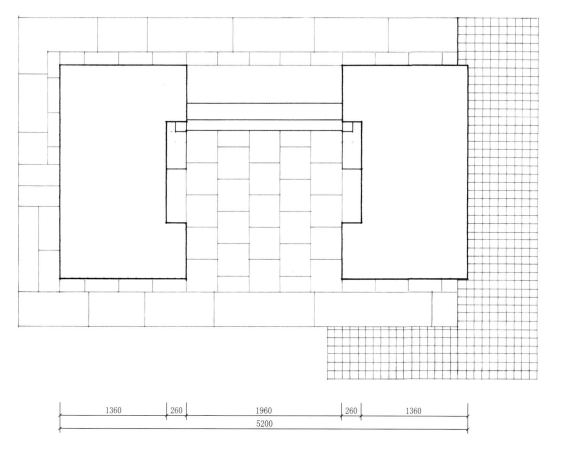

社稷坛宰牲亭旁门平面图
Plan of the Side entrance of the Butcher Pavilion of the Altar of Land and Grain

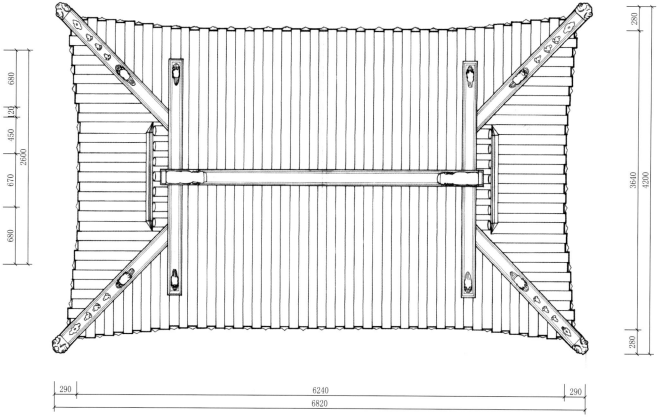

社稷坛宰牲亭旁门屋顶平面图
Roof plan of the Side Entrance of the Butcher Pavilion of the Altar of Land and Grain

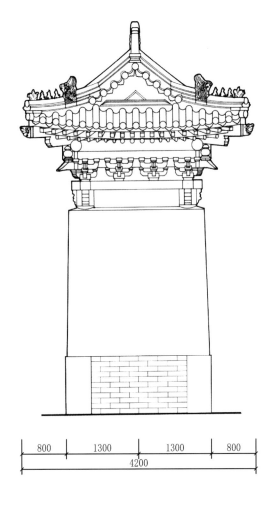

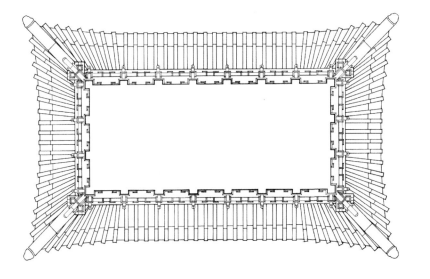

800　　1300　　1300　　800
4200

社稷坛宰牲亭旁门侧立面图
Side elevation of the Side entrance of the Butcher Pavilion of the Altar of Land and Grain

社稷坛宰牲亭旁门屋顶仰视平面图
Bottom plan of the Side entrance of the Butcher Pavilion of the Altar of Land and Grain

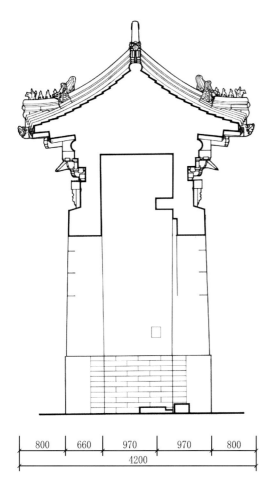

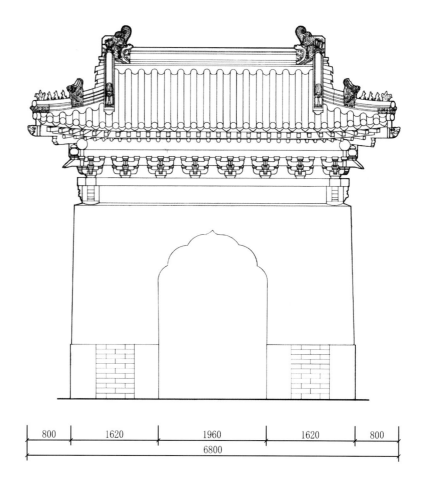

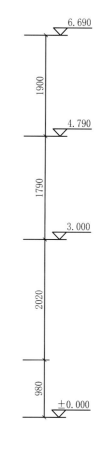

社稷坛宰牲亭旁门横剖面图
Cross section of the Side entrance of the Butcher Pavilion of the Altar of Land and Grain

社稷坛宰牲亭旁门正立面图
Front elevation of the Side entrance of the Butcher Pavilion of the Altar of Land and Grain

主要参考文献

[一] (明) 明太祖实录 [M]. 台北：[中央研究院] 历史语言研究所校勘影印，1962.

[二] (明) 明英宗实录 [M]. 台北：[中央研究院] 历史语言研究所校勘影印，1962.

[三] (明) 明武宗实录 [M]. 台北：[中央研究院] 历史语言研究所校勘影印，1962.

[四] (明) 明世宗实录 [M]. 台北：[中央研究院] 历史语言研究所校勘影印，1962.

[五] (明) 永乐大典 [M]. 北京：中华书局，1962.

[六] (明) 申时行. 明会典 [M]. 北京：中华书局，1989.

[七] (明) 王俊华. 洪武京城图志 [M]. 北京：书目文献出版社，1998.

[八] (明) 李贤 等. 明一统志.

[九] (明) 徐一夔 等. 大明集礼. 日本早稻田大学藏书，明嘉靖本.

[十] (清) 清朝通典 [M]. 台北：新兴书局，1958.

[十一] (清) 清高宗纯皇帝实录 [M]. 北京：中华书局，1985.

[十二] (清) 昆冈 等. 钦定大清会典图 (影印清光绪石印本) [M]. 北京：书目文献出版社，1998.

[十三] (清) 允裪 等. 钦定大清会典则例 (文渊阁四库全书电子版). 上海：上海出版社，迪志文化出版有限公司，1999.

[十四] (清) 伊桑阿 等. 康熙朝钦定大清会典 [M] // 近代中国史料丛刊三编. 第72辑. 台北：文海出版社有限公司，1992.

[十五] (清) 托津 等. 嘉庆朝钦定大清会典事例 [M] // 近代中国史料丛刊三编. 第69辑 (卷584-725). 台北：文海出版社有限公司，1992.

[十六] (清) 托津 等. 嘉庆朝钦定大清会典图 [M] // 近代中国史料丛刊三编. 第71辑. 台北：文海出版社有限公司，1992.

[十七] (清) 孙承泽. 春明梦余录 [M]. 北京：北京古籍出版社，1992.

[十八] (清) 于敏中. 日下旧闻考 [M]. 北京：北京古籍出版社，1983.

[十九] (清) 张廷玉. 明史 [M]. 北京：中华书局，1976.

[二十] 曹鹏. 明代都城坛庙建筑研究 [D]. 天津：天津大学博士学位论文，2011.

[二十一] 池小燕，曹鹏. 北京地坛与社稷坛祭祀对象之辨——略述地祇神与社稷神各历史时期发展演变 [J]. 沈阳建筑大学学报 (社会科学版)，2006(4)：309-312.

[二十二] 傅公钺. 清代的太庙 [J]. 故宫博物院院刊，1986(3)：73-79+86.

[二十三] 郭华瑜. 明代官式建筑大木作研究 [M]. 南京：东南大学出版社，2005.

[二十四] 姜舜源. 清代的宗庙制度 [J]. 故宫博物院院刊，1987(3)：15-23+57.

[二十五] 孙大章. 中国建筑艺术全集 9·坛庙建筑 [M]. 北京：中国建筑工业出版社，2000.

[二十六] 天津大学建筑设计研究院，北京市劳动人民文化宫. 全国重点文物保护单位北京太庙保护规划. 2014.

[二十七] 亚白杨. 北京社稷坛建筑研究 [D]. 天津：天津大学硕士学位论文，2005.

[二十八] 亚白杨. 明清社稷坛空间布局设计意向——「左祖右社」格局探源 [J]. 古建园林技术，2010(2)：57-59.

[二十九] 闫凯，王其亨，曹鹏. 北京明清皇家三大殿之比较研究 [J]. 山东建筑工程学院学报，2006(2)：116-121.

[三十] 闫凯. 北京太庙建筑研究 [D]. 天津：天津大学硕士学位论文，2004.

[三十一] 中山公园管理处. 中山公园志 [M]. 北京：中国林业出版社，2002.

# References

(1) (Ming) Chronicles of Taizu of Ming Dynasty. Taipei: Emendated photocopy by Institute of History and Philology, Academia Sinica, 1962;

(2) (Ming) Chronicles of Yingzong of Ming Dynasty. Taipei: Emendated photocopy by Institute of History and Philology, Academia Sinica, 1962;

(3) (Ming) Chronicles of Wuzong of Ming Dynasty. Taipei: Emendated photocopy by Institute of History and Philology, Academia Sinica, 1962;

(4) (Ming) Chronicles of Shizong of Ming Dynasty. Taipei: Emendated photocopy by Institute of History and Philology, Academia Sinica, 1962;

(5) (Ming) Great Encyclopedia of the Yongle Reign. Beijing: Zhonghua Book Company, 1962;

(6) (Ming) Shen Shixing. Collected Statutes of the Ming Dynasty. Beijing: Zhonghua Book Company, 1989;

(7) (Ming) WANG Junhua. Hongwu Capital City Maps. Beijing: Bibliography and Document Publishing House, 1998;

(8) (Ming) LI Xian, et al. Total Annals of the Ming Dynasty. Full-text Video Database of Chinese Rare Books;

(9) (Ming) XU Yikui, et al. A Collection of Etiquettes of the Ming Dynasty. held in Waseda University, Japan, edition of the Jiajing Period of the Ming Dynasty;

(10) (Qing) Encyclopedia of Qing Dynasty. Taipei: Xinxing Book Company. 1958;

(11) (Qing) Chronicles of Gaozong of Qing Dynasty. Beijing: Zhonghua Book Company, 1985.

(12) (Qing) Kun Gang, et al. Emperor Authorized Collected Statutes of the Great Qing Dynasty. Lithography photocopy of Emperor Guangxi of the Qing Dynasty. Taipei: Qiwen Publishing House, 2006;

(13) (Qing) Yun Tao, et al. Emperor Authorized Code of Collected Statutes of the Great Qing Dynasty. Electronic edition of the Complete Library in Four Branches of Literature. Shanghai: Shanghai Publishing House, Digital Heritage Publishing Ltd., 1999;

(14) (Qing) Isangga et al. Collected Statutes of the Great Qing Dynasty (compiled in the Kangxi Period). Edition 3 of A Series of Historical Materials of Modern China, Part 72. Taipei: Wenhai Publishing House Co., Ltd, 1992;

(15) (Qing) Tuojin et al. Emperor Authorized Cases of Collected Statutes of the Great Qing Dynasty (compiled in the Jiaqing Period). dition 3 of A Series of Historical Materials of Modern China, Part 69 (Volume 584-725). Taipei: Wenhai Publishing House Co., Ltd, 1992;

(16) (Qing) Tuojin et al. Emperor Authorized Graphs of Collected Statutes of the Great Qing Dynasty (compiled in the Kangxi Period). Edition 3 of A Series of Historical Materials of Modern China, Part 71. Taipei: Wenhai Publishing House Co., Ltd, 1992;

(17) (Qing) Sun Chengze. Local Chronicles of Beijing in the Ming Dynasty. Beijing: Beijing Ancient Works Publishing House, 1992;

(18) (Qing) Yu Minzhong. Research on Archived News of Qing Capital. Beijing: Beijing Ancient Works Publishing House, 1983;

(19) (Qing) Zhang Tingyu. History of Ming Dynasty. Beijing: Zhonghua Book Company, 1976;

(20) Cao Peng. Study on Temple and Altar Architecture in the Ming Dynasty. Tianjin: doctoral dissertation of Tianjin University, 2011;

(21) Chi Xiaoyan, Cao Peng. Discrimination of Targets of Worship of Beijing Temple of Earth and Altar of Land and Grain - Outline on Evolution of Gods of Earth as well as Land and Grain Through Historical Periods. Journal of Shenyang Jianzhu University (Social Science Edition), 2006(4), 309-312;

(22) Fu Gongyue. Ancestral Temple in the Qing Dynasty. Journal of the Palace Museum, 1986(3): 73-79+86;

(23) Guo Huayu. Study on Carpentry Work of Official Architecture in the Ming Dynasty. Nanjing: Southeast University Press, 2005;

(24) Jiang Shunyuan. Ancestral Temple System in the Qing Dynasty. Journal of the Palace Museum, 1987(3). PP. 15-23+57;

(25) Sun Dazhang. Complete Works of Chinese Architectural Art Part 9 · Altars and Shrines Buildings. Beijing: China Architecture & Building Press, 2000;

(26) Architectural Design and Research Institute of Tianjin University, Beijing Working People's Cultural Palace. Protection Planning of Ancestral Temple as a key cultural relics unit under national protection, 2014;

(27) Ya Baiyang. Research on Architecture of Beijing Altar of Land and Grain. Tianjin: master dissertation of Tianjin University, 2005;

(28) Ya Baiyang. Intent of Space Layout Design of the Altar of Land and Grain in Ming and Qing Dynasties - Exploring the Origin of the "temple in east and altar in west" Layout. Traditional Chinese Architecture and Gardens, 2010(2). PP. 57-59;

(29) Yan Kai, Wang Qiheng, Cao Peng. Comparative Study on the Three Royal Palaces of the Ming and Qing Dynasties in Beijing. Journal of Shandong Institute of Architecture & Engineering, 2006(2). PP. 116-121;

(30) Yan Kai. Research on the Architecture of Imperial Ancestral Temple in Beijing. Tianjin: master dissertation of Tianjin University, 2004;

(31) Beijing Zhongshan Park Administration. Chronicle of Beijing Zhongshan Park. Beijing: China Forestry Press, 2002.

# 参与测绘及相关工作的人员名单

## 1997 年太庙测绘

指导教师：王其亨 刘彤彤 何捷 张晓宇 白雪海

指导研究生：李海涛

测绘学生：陈恭锦 陈津 段建立 丁利群 风清扬 付东楠
宫媛 郭力 贺军 李佳 李迈 李巍
廖超 林勇强 刘海 刘力飞 苗展堂 沈粤
史逸 王虹 王晶 王茜 王绚 王臻倬
谢云 熊巍 徐茂臻 杨晓龙 杨轶 叶沧海
叶楠 叶强 于雷 于志渊 袁海滨 张丹琪
张萍 赵杨 郑可佳 齐洪海 赵侠

## 1997 年社稷坛测绘

指导教师：王其亨 赵建波 白雪海 荆子洋 盛海涛 窦征
杨益纪业

指导研究生：李海涛

测绘学生：王飞 王宇 王虹 王绚 王晶
王雪莲
风清扬 方铁英 石媛媛 叶强 叶楠 史逸
吕强 朱俊 刘芳 刘恒 孙泽山 孙潮蔚
李业颖 李迈 李佳 杨梅 杨猛 汪琦
沈粤 张萍 张惠青 陈扬 陈津 陈鹏
林勇强 庞芳芳 宗菲 赵扬 胡莺
施强 宫媛 贺军 秦彦波 袁海滨 钱志嵩
郭力 董亚图 程鹏 谢云 慎小嶷 廖超
李杨 李梓华 王科 崔萍

1998 年太庙测绘

指导教师：王其亨　刘彤彤　何捷

本科生：曹鹏　杨梅

测绘学生：戈斌　付斌　邓烨　耿武　李靳　林源
闫凯　李天骄　缪波　苏薇　张蕾　汤岳
邢雪莹　郑宁　魏星　刘巍　刘晓雪　王晶
荆锋　齐中凯　曾鹏　高云朗　胡志欣　黄澜
廉洁　刘培岩　柳岩　施华　王岩　王振飞
王智　杨军　游亚鹏　于爽　张汀　赵晓刚

2014 年太庙保护规划

指导教师：王其亨

博士研究生：袁守愚　刘翔宇

本科生：邓晓琳　朱英宁　荆国栋　金银实　辛鹏飞　李超

2017 年、2021 年太庙和社稷坛测绘图纸整理及翻译

指导教师：王其亨　何捷　张凤梧

博士研究生：曹睿原　马昭仪　王磊

硕士研究生：王禹　牛嘉诚　勾乐梅　李琦　雷巍　刘帅帅
张帅　张舒　杨明　沈孙乐　李峰　袁诗雨

英文翻译：金田　尚晋

# List of Participants Involved in Surveying and Related Works

**The Ancestral Temple Surveying and Mapping in 1997**

Instructor: WANG Qiheng, LIU Tongtong, HE Jie, ZHANG Xiaoyu, BAI Xuehai

Graduate Student: LI Haitao

Student: CHEN Gongjin, CHEN Jin, DUAN Jianli, DING Liqun, FENG Qingyang, FU Dongnan, GONG Yuan,GUO Li, HE Jun, LI Jia, LI Mai, LI Wei, LIAO Chao, LIN Yongqiang, LIU Hai, LIU Lifei, MIAO Zhantang, SHEN Yue, SHI Yi, WANG Hong, WANG Jing, WANG Qian, WANG Xuan, WANG Zhenzhuo, XIE Yun, XIONG Wei, XU Maozhen, YANG Xiaolong, YANG Yi, YE Canghai, YE Nan, YE Qiang, YU Lei, YU Zhiyuan, YUAN Haibin, ZHANG Danqi, ZHANG Ping, ZHAO Yang, ZHENG Kejia, QI Honghai, ZHAO Xia

**The Altar of Land and Grain Surveying and Mapping in 1997**

Instructor: WANG Qiheng, ZHAO Jianbo, BAI Xuehai, JING Ziyang, SHENG Haitao, DOU Zheng, YANG Yi, JI Ye

Graduate Student: LI Haitao

Student: WANG Fei, WANG Yu, WANG Hong, WANG Xuan, WANG Xuelian, WANG Jing, FENG Qingyang, FANG Tieying, SHI Yuanyuan, YE Qiang, YE Nan, SHI Yi, LV Qiang, ZHU Jun, LIU Fang, LIU Heng, SUN Zeshan, SUN Chaowei, LI Yeying, LI Mai, LI Jia, YANG Mei, YANG Meng, WANG Qi, SHEN Yue, ZHANG Ping, ZHANG Huiqing, CHEN Yang, CHEN Jin, CHEN Peng, LIN Yongqiang, PANG Fangfang, ZONG Fei, ZHAO Feiqi, ZHAO Yang, HU Ying, SHI Qiang, GONG Yuan, HE Jun, QIN Yanbo, YUAN Haibin, QIAN Zhisong, GUO Li, DONG Yatu, CHENG Peng, XIE Yun, SHEN Xiaoyi, LIAO Chao, LI Yang, LI Zihua, WANG Ke, CUI Ping

**The Ancestral Temple Surveying and Mapping in 1998**

Instructor: WANG Qiheng, LIU Tongtong, HE Jie

Bachelor Student: CAO Peng, YANG Mei

Student: GE Bin, FU Bin, DENG Ye, GENG Wu, LI Jin, LIN Yuan, YAN Kai, LI Tianjiao, MOU Bo, SU Wei, ZHANG Lei, TANG Yue, XING Xueying, ZHENG Ning, WEI Xing, LIU Wei, LIU Xiaoxue, WANG Jing, JING Feng, QI Zhongkai, CENG Peng, GAO Yunlang, HU Zhixin, HUANG Lan, LIAN Jie, LIU Peiyan, LIU Yan, SHI Hua, WANG Yan, WANG Zhenfei, WANG Zhi, YANG Jun, YOU Yapeng, YU Shuang, ZHANG Ting, ZHAO Xiaogang

**The Ancestral Temple Protection Planning in 2014**

Instructor: WANG Qiheng

Doctoral Student: YUAN Shouyu, LIU Xiangyu

Bachelor Student: DENG Xiaolin, ZHU Yingning, JING Guodong, JIN Yinshi, XIN Pengfei, LI Chao

**The Ancestral Temple and the Altar of Land and Grain Drawings and Translation in 2017and 2021**

Instructor: WANG Qiheng, HE Jie, ZHANG Fengwu

Doctoral Student: CAO Ruiyuan, MA Zhaoyi, WANG Lei

Master Student: WANG Yu, NIU Jiacheng, GOU Lemei, LI Qi, LEI Wei, LIU Shuaishuai, ZHANG Shuai, ZHANG Shu, YANG Ming, SHEN Sunle, LI Feng YUAN Shiyu

English Translator: Aurelia Campbell, SHANG Jin

**图书在版编目（CIP）数据**

太庙和社稷坛 = THE ANCESTRAL TEMPLE AND THE ALTAR OF LAND AND GRAIN：汉英对照 / 王其亨主编；何捷，王其亨编著；天津大学建筑学院，北京市劳动人民文化宫，北京市中山公园管理处合作编写 . — 北京：中国建筑工业出版社，2019.12

（中国古建筑测绘大系 . 坛庙建筑）

ISBN 978-7-112-24548-2

Ⅰ.①太… Ⅱ.①王… ②何… ③天… ④北… ⑤北 …Ⅲ.①寺庙—宗教建筑—建筑艺术—北京—图集 Ⅳ. ① TU-885

中国版本图书馆CIP数据核字（2019）第284485号

丛书策划 / 王莉慧

责任编辑 / 李　鸽　刘　川

英文翻译 / 金　田　尚　晋

书籍设计 / 付金红

责任校对 / 王　烨

中国古建筑测绘大系 · 坛庙建筑

**太庙和社稷坛**

天 津 大 学 建 筑 学 院

北京市劳动人民文化宫　　合作编写

北京市中山公园管理处

何捷　王其亨　编著　王其亨　主编

Traditional Chinese Architecture Surveying and Mapping Series: Temples Architecture

**THE ANCESTRAL TEMPLE AND THE ALTAR OF LAND AND GRAIN**

Compiled by School of Architecture, Tianjin University

Beijing Working People's Culture Palace

Administration Office of Beijing Zhongshan Park

Chief Edited by WANG Qiheng Edited by HE Jie, WANG Qiheng

\*

中国建筑工业出版社出版、发行（北京海淀三里河路9号）

各地新华书店、建筑书店经销

北京方舟正佳图文设计有限公司制版

北京雅昌艺术印刷有限公司印刷

\*

开本：787毫米×1092毫米　横 1/8　印张：28½　字数：716千字

2022年2月第一版　2022年2月第一次印刷

定价：**238.00** 元

ISBN 978-7-112-24548-2

（35216）